Tomorrow's Energy

Tomorrow's Energy
Hydrogen, Fuel Cells, and the Prospects for a Cleaner Planet

Peter Hoffmann

The MIT Press
Cambridge, Massachusetts
London, England

First MIT Press paperback edition, 2002

Set in Sabon by The MIT Press.
Printed and bound in the United States of America.

Library of Congress Cataloging-in-Publication Data

Hoffmann, Peter, 1935–
Tomorrow's energy : hydrogen, fuel cells, and the prospects for a cleaner planet / Peter Hoffmann.
p. cm.
Includes bibliographical references and index.
ISBN 0-262-08295-0 (hc. : alk. paper), 0-262-58221-X (pb)
1. Hydrogen as fuel. I. Title.

TP359.H8 H633   2001
333.79'4—dc21   00-054613

10 9 8 7 6

Contents

Foreword

Senator Tom Harkin

When I was born, in 1939, there were 2 billion people on the earth. When I turned 60, there were 6 billion. When my daughter turns 60, there will be 9 billion. Many of these people will want heating in the winter and air conditioning in the summer. They will want to use refrigerators, radios, televisions, and cars.

The question is not whether nations like China and India will develop or whether they will consume more energy than they do now. They will and they should. They have a right to seek a better life just as we did.

The question is: What kind of world will we create? If we continue to base our economies on coal and oil, we will create a world with toxic air, filthy water, and debilitating diseases. Global warming will likely bring droughts and hurricanes, tropical diseases in the North, and widespread coastal flooding.

There is an alternative. We can replace coal and oil with clean, renewable energy sources that can generate electricity, heat buildings, and power cars.

Renewable energy sources are abundant throughout the world. India is flooded with sunlight, and China's entire current electricity consumption could be powered by the wind in Inner Mongolia. In the United States, the Midwest is sometimes called the Saudi Arabia of wind.

However, we must remember that solar, wind, and most other renewable energy sources are intermittent and regional. They can only become major power sources if we find a way to store and transport their energy efficiently.

Hydrogen can make the renewable vision real by storing renewable energy and making it available where and when it is needed. Hydrogen, the simplest molecule, is non-toxic and can be made from plain water using electricity from renewable sources. Used in fuel cells, hydrogen generates electricity and emits only water vapor. And cars run on hydrogen fuel cells are 2 to 3 times more efficient than gasoline engines. Fuel cells can be made in any size to fit everything from pocket-held devices to large power plants. They are perfect for a dispersed and robust energy infrastructure.

This book is the culmination of Peter Hoffmann's work over the past three decades to chronicle the progress of hydrogen energy from a vision to a niche market to its position today on the brink of full commercialization. He describes the various ways hydrogen can be made, stored, and used, and offers insightful analyses of the remaining technical and economic obstacles to the widespread use of hydrogen.

Throughout my career in the Senate, I have worked to promote the development of a hydrogen economy. I'm glad to say I've seen tremendous progress. Today there are hydrogen fueling stations and fuel cell buses scattered in cities around the world. DaimlerChrysler intends to sell fuel cell cars commercially by 2004, with other automakers close behind. The First National Bank of Omaha is using fuel cells to provide secure power for its credit card service operations. And recently, the New York City Police Department installed a fuel cell to power its Central Park Police Station because it was cheaper than extending power lines.

Hoffmann's vast knowledge and insight on hydrogen will be an invaluable tool for continuing these efforts, and an important resource for anyone who cares about our environment. After reading this book, one can see the dream of a hydrogen-based economy becoming a reality. I am confident that I will one day walk from my hydrogen-heated office through clean air to my hydrogen fuel cell car. And when I do, I will be carrying this book.

Acknowledgments

This book started out as a revision and an update of *The Forever Fuel—The Story of Hydrogen*, published by the Westview Press in 1981. I hadn't really intended to ever put myself through the wringer of writing a book again, but I gave in to the entreaties of a few people who thought such a book was needed because hydrogen and fuel cell energy have in fact made progress since *The Forever Fuel* and because a new and improved version might help accelerate the process a bit more. As I began gathering new material (aided greatly by having to report each month for *The Hydrogen & Fuel Cell Letter*, which I and my wife, Sarah, had started in 1986), the book just sort of grew beyond a revised edition. The MIT Press and I think it's a new book, although quite a bit of the historical material has been retained in condensed form.

Thanks are due to many people for all sorts of reasons—providing information and input, suggesting new areas, reading early parts critically, suggesting changes, offering moral support when I was ready to chuck it all. Some of them are mentioned in the book. They include, in no particular order, Alan Lloyd, Shannon Baxter, Sandy Thomas, Jesse Ausubel, Henry Linden, Dan Brewer, Gary Sandrock, Joan Ogden, Bob Williams, Bob Zweig, T. Nejat Veziroglu, Bob Rose, Frank Lynch, Karl Kordesch, Peter Lehman, Oliver Weinmann, James Provenzano, Chris Borroni-Bird, Venki Raman, Firoz Rasul, Neil Otto, Debby Harris, Carl-Jochen Winter, Ulrich Schmidtchen, Ron Sims, Cesare Marchetti, Hjalmar Arnason, Heinz Klug, Hans Pohl, Reiner Wurster, Ulrich Buenger, Vahe Kludjan, Martin Hammerli, Karen Miller, Bob Mauro, Lowell Ungar, Ranji George, Barbara Heydorn, Olof Tegström, Curtis Moore, Marcus Nurdin, John Turner,

Paul Weaver, Sandy and Andrew Stuart, Glenn Rambach, James Cannon, John O'Sullivan, Jeff Bentley, and Dr. Ulrike Gutzmann.

Special thanks for helping me obtain financial assistance are due to Bill Hoagland, Susan Leach, Neil Rossmeissl, and Cathy Gregoire Padro of the US Department of Energy.

Finally, I want to thank Clay Morgan and Paul Bethge of The MIT Press, my wife, Sarah, and Taylor M. Briggs, who spent untold hours checking copy, making suggestions, removing some of the more egregious errors, and in general helping to pound the manuscript into shape. Without them, there would have been no book.

Tomorrow's Energy

1

Why Hydrogen? Buckminster Fuller, Sheikh Yamani, and Bill Clinton

There are two prime sources of energy to be harnessed and expended to do work. One is the capital energy-saving and storage account; the other is the energy-income account. The fossil fuels took multimillions of years of complex reduction and conservation, progressing from vegetational impoundment of sun radiation by photosynthesis to deep-well storage of the energy concentrated below the earth's surface. There is a vast overabundance of income energy at more places around the world, at more times to produce billionsfold the energy now employed by man, if he only knew how to store it when it is available, for use when it was not available. There are gargantuan energy-income sources available which do not stay the processes of nature's own conservation of energy within the earth's crust "against a rainy day." These are in water, tidal, wind, and desert-impinging sun radiation power. The exploiters of the fossil fuels, coal and oil, say it costs less to produce and burn the savings account, This is analogous to saying it takes less effort to rob a bank than to do the work which the money deposited in the bank represents. The question is cost to whom? To our great-great-grandchildren who will have no fossil fuels to turn the machines? I find that the ignorant acceptance by world society's presently deputized leaders of the momentarily expedient and the lack of constructive, long-distance thinking—let alone comprehensive thinking—would render dubious the case for humanity's earthian future could we not recognize plausible overriding trends.

—R. Buckminster Fuller, 1969[1]

The big powers are seriously trying to find alternatives to oil by seeking to draw energy from the sun or water. We hope to God they will not succeed quickly because our position in that case will be painful.

—Sheikh Ahmad Zaki Yamani, oil minister of Saudi Arabia, 1976[2]

A shift to solar energy could create 2.9 million jobs and cut spending for conventional fuels by $11.8 billion by 1990, according to a study released yesterday by Sen. Edward Kennedy, D-Mass.

—United Press International, April 22, 1979

Hydrogen as fuel? It's still Buck Rogers stuff.

—energy expert, Bonn, February 1980

Ballard Power and United Technologies are leading pioneers in developing fuel cells that are so clean. Their only exhaust is distilled water. Right now, Ballard is working with Chrysler, Mercedes-Benz and Toyota to introduce fuel cells into new cars.
—President Bill Clinton, 1997[3]

In the twenty-first century hydrogen might become an energy carrier of importance comparable to electricity. This is a very important mid- to long-range research area.
—President's Committee of Advisors on Science and Technology, 1997[4]

We believe that hydrogen fuel cell powered cars are likely to make a major entrance into the vehicle market throughout Europe and the US by 2005. In addition, we see potentially enormous opportunities opening up in the domestic fleet, bus and taxi market as government encourages cleaner alternatives to conventionally powered vehicles. This trend poses a real challenge to a company like Shell to develop new products, new technologies and to prepare and inform our customers for the changes that lie ahead. . . . I can assure you we are in this for the long haul.
—Chris Fay, chief executive, Shell UK, London, 1998[5]

Our long-term goal is very simple: zero emissions in the air. To meet that goal, California has teamed with some of the best automotive manufacturers and energy providers in the world to develop an exciting new technology that is both environmentally safe and commercially viable.
—Gray Davis, Governor of California, 1999[6]

Now analysts say that natural gas, lighter still in carbon, may be entering its heyday, and that the day of hydrogen—providing a fuel with no carbon at all, by definition—may at last be about to dawn.
—*New York Times*, 1999[7]

These quotes give some idea as to what this book is all about: hydrogen as a non-polluting, renewable form of energy. Hydrogen—an invisible, tasteless, colorless gas—is the most abundant element in the universe. It is the fuel of stars and galaxies. Highly reactive, it is essential in innumerable chemical and biological processes. It is an energetic yet (by definition) non-polluting[8] fuel.

Even before Buckminster Fuller's observations, many people had been calling for the use of nature's "current energy account" (solar power in its various manifestations) as an alternative to robbing the world's energy "savings account" (coal, oil, gas). As Fuller pointed out, the problem has been to a large extent not only how to collect this essentially free energy but how to store it. Tapping into solar energy for purposes other than basic solar heating usually means producing electricity. But electricity has to be consumed the instant it is produced. It is difficult to store in large quantities. Hydrogen, a storable gas, solves that problem.

In past decades, efforts to harness renewable energies were driven partly by idealism but more by concerns about "energy security"—fears about the eventual drying up of the world's petroleum resources and about the increasing vulnerability of the long supply lines from the politically unstable Middle East. But as the twentieth century drew to its close, environmental concern had become a much stronger impetus driving the world toward renewable, alternative forms of energy. Curbing and eventually doing away with pollution has become a universal concern. Dying forests in Europe and acid rain everywhere were among the initial wake-up calls to the need to curb sulfur, nitrogen oxides, hydrofluorocarbons (HFCs), perfluorocarbons (PFCs), particulate emissions, and other pollutants. At last it had begun to dawn on policy makers and large parts of the general population—less so, and more slowly, in the United States than in other parts of the world—that the very process of combusting fossil fuels, the interaction of carbon in hydrocarbon fuels with the air's oxygen, and the consequent release into and accumulation in the atmosphere of carbon dioxide, carbon monoxide, and other climate-changing gases far above pre-industrial levels was raising the world's temperature—the famous Greenhouse Effect—and threatening to play havoc with the world's climate.

"Zero emissions" from cars and buses, industry, ships, and home furnaces is becoming the new world standard—a standard to which industrialized countries and emerging economies are aspiring with varying degrees of intensity and dedication. To the minds of many, taking the carbon out of hydrocarbons and relying on the "hydro" part—hydrogen—as a zero-emission chemical fuel is the obvious though technically difficult way to minimize and, it is hoped, eventually eliminate global warming.

The basics of global warming are as follows: Carbon dioxide (CO_2) is produced by the burning of fossil fuels as well as by nature's carbon cycle. (Humans and animals exhale it into the atmosphere as part of their metabolic process; green plants absorb it and turn it into plant matter.) CO_2, methane, and other gases act like a greenhouse in the atmosphere: They let solar radiation through the atmosphere to heat the Earth's surface, but they prevent the reradiation of some of that energy back into space, thus trapping heat. Some heat entrapment is good; otherwise we would have never evolved in the first place, or we would freeze to death. But the more greenhouse gases are swirling around the atmosphere, the more heat is trapped. Because of decreases in the world's forests and consequent decreases in global CO_2 absorption, and (more important) because of increasing burning of fossil fuels in our ever-more-energy-demanding machinery, the atmosphere's CO_2 content has been going up steadily and increasingly steeply since the beginning of the Industrial Revolution.

Aside from other fundamental climate cycles stretching over thousands or tens of thousands of years (such as ice ages, believed to be caused in parts by changes in sunspots and therefore beyond man's ability to influence), Earth's climate has been reasonably stable for 10,000 years or so. But this equilibrium is being upset by man-made carbon emissions. The question is how much. Opinions, basic assumptions about the future course of the climate and the amount of expected heat increase, closely related assumptions about global economic development, and faith in the complex computer models that attempt to forecast climate developments vary widely even among the majority of experts who believe that our planet is facing an unprecedented crisis.[9]

As more heat is being trapped, and as temperatures climb all over the world, the mainstream opinion among the climate experts of the United Nations' Intergovernmental Panel on Climate Change (IPCC) predicts widespread and drastic impacts on ecosystems, water resources, food and fiber production, coastlines, and human health: The polar ice caps will melt, sea levels will rise, large stretches of coastline (including some of the world's biggest cities) will be inundated, and scores of islands in the Pacific may disappear. Agricultural patterns are likely to change, with grain-growing belts migrating northward. The middle to high latitudes may become more productive as plants absorb more available CO_2. The agricultural yields of the tropics and the subtropics are expected to decrease.

Climate change could produce more deaths through heat stress, the spread of tropical diseases, and worse urban air pollution. In a special supplement dedicated to the December 1997 Global Climate Change conference in Kyoto, the *New York Times* reported that one IPCC working group had summarized its findings as follows: "Compared with the total burden of ill health, these problems are not likely to be large. In the aggregate, however, the direct and indirect impacts of climate change on human health do constitute a hazard to human population health, especially in the developing countries in the tropics and subtropics."[10]

A recent study[11] that looked at the generation of ozone in four metropolitan areas (Sacramento, Chicago, St. Louis, and Los Angeles) concluded that a future doubling of global atmospheric CO_2 would likely result in higher daily temperatures, which in turn would "dominate the meteorological correlations with high tropospheric ozone concentrations"—in other words, higher temperatures would increase the ozone concentrations. More ozone, in turn, would increase the incidence of premature mortality, hospital admissions for respiratory diseases, and respiratory symptoms, the authors said. But some aspects, especially the relationship between ozone levels and premature mortality, are still subject to ongoing research, one author cautioned. In the case of Los Angeles, doubled CO_2 concentrations were expected to increase the annual average daily maximum temperature from the base-case 20.7°C to 24.9°C and the annual average daily minimum from 14.1°C to 18.2°C, the researchers calculated. In Chicago, doubled CO_2 would increase the corresponding maximum from 13.5°C to 19.3°C and the minimum from 3.78°C to 10.0°C. For Los Angeles, a table of anticipated extra health costs for one such warmer future year listed $2.552 billion (in 1990 dollars) for premature mortality, $14.19 million for hospital admissions, and $168,000 for respiratory-symptom days relative to the same cost categories for a typical recent year. For Chicago, the corresponding numbers were $979 million, $2.38 million, and $28,000.

The other principal form of clean energy, electricity, has two strikes against it: (1) It is the minority component in the world's energy production and consumption—chemical energy accounts for almost two-thirds. (2) Most electricity is produced by burning fossil fuels—coal, natural gas, petroleum. According to the 1997 edition of the US Department of Energy's International Energy Outlook, the world's total energy consumption in 1995 was close to 364 quads (quadrillion British thermal units).[12] Of that, 140 quads (38 percent) consisted of electricity. Of that electricity, 62 percent was produced by burning oil, coal, or gas. (Coal accounted for the biggest slice—51.6 quads.) Renewable energy—mostly hydroelectric—accounted for only 29.7 quads, and nuclear energy only 23.3 quads. Oil, widely seen as the root cause of our energy woes because of its familiarity as fuel for our vehicles and because of the periodic political antics of the Middle Eastern countries and the Organization of Petroleum Exporting Countries, actually accounted for the smallest share of the world's energy consumption, with 12.9 quads.

Thus, it is safe to say that, in general, we work and play with—and, environmentalists would say, more frequently than ever die from—fossil-fueled chemical energy. Gasoline, diesel fuel, heavy oil, jet-grade kerosene, natural gas, wood, biomass, and coal propel airplanes, cars, trains and ships, run plants, and heat homes, offices, hospitals, and schools. Hydrogen, also a form of chemical energy, can do all those things, and can do them essentially without polluting.

When burned in an internal-combustion engine (piston, rotary, or gas turbine), hydrogen produces as exhaust virtually nothing but harmless water vapor (plus, admittedly, trace emissions from tiny amounts of engine lubricants that are oxidized in the process, and some nitrogen oxides).[13] When hydrogen is combusted with atmospheric oxygen in an engine, no carbon monoxide or carbon dioxide is emitted, no unburned hydrocarbons, no stench, no smoke, nor any of the other carbon-bearing, Earth-befouling discharges we suffer today.

Hydrogen performs even better in *fuel cells* (electrochemical engines that, by electrochemically combining hydrogen and oxygen in a flameless process, produce electricity, heat, and pure, distilled water—the mirror image of electrolysis, in which water is split into hydrogen and oxygen by running a current through it). Unlike internal-combustion engines, fuel cells produce no nitrogen oxides at all.[14]

Fuel cells have no moving parts. Nearly silent, they can be as much as 2.5 times as efficient as internal-combustion engines. In the 1990s they became widely and publicly recognized as a vanguard technology that may launch hydrogen energy on its way to becoming a major, environmentally benign, sustainable, renewable component of the world's energy mix for both transportation and stationary applications.

"Hydrogen, H_2, atomic weight 1.00797 . . . is the lightest known substance," reports the *Encyclopedia of Chemistry*. "The spectroscope shows that it is present in the sun, many stars, and nebulae. Our galaxy . . . plus the stars of the Milky Way is presently considered to have been formed 12 to 15 billion years ago from a rotating mass of hydrogen gas which condensed into stars under gravitational forces. This condensation produced high temperatures, giving rise to the fusion reaction converting hydrogen into helium, as presently occurring in the sun, with the evolution of tremendous amounts of radiant thermal energy plus the formation of the heavier elements. Hydrogen gas has long since escaped from the Earth's lower atmosphere but is still present in the atmosphere of several of the planets. In a combined state, hydrogen comprises 11.19 percent of water and is an essential constituent of all acids, hydrocarbons, and vegetable and animal matter. It is present in most organic compounds."[15]

Hydrogen is used in many industries as a chemical raw material, especially in the production of fertilizer, but also in making dyes, drugs, and plastics. It is used in the treatment of oils and fats, as a fuel for welding, to make gasoline from coal, and to produce methanol. In its supercold liquid form, in combination with liquid oxygen, it is a powerful fuel for the Space Shuttle and other rockets.

Hydrogen is produced commercially in almost a dozen processes. Most of them involve the extraction of the "hydro" part from hydrocarbons. The most widely used, least costly process is "steam reforming," in which natural gas is made to react with steam, releasing hydrogen. Water electrolysis, in which water is broken down into hydrogen and oxygen by running an electrical current through it, is used where electricity is cheap and where high purity is required.

Hydrogen can be stored as a high-pressure gas or as an integral component in certain alloys known as hydrides, but also (a recent development) in and on microscopic carbon fibers. As a cryogenic liquid fuel, it promises to lead to better, faster, more efficient, environmentally "clean" airplane designs. Metallic hydrogen, a laboratory curiosity so far, holds promise as an ultra-energetic fuel and as a zero-resistance electrical conductor in all sorts of electrical and electronic technologies.

Since the 1930s, environment-minded scientists, academics, and energy planners (inside and outside government), industrial executives, and even some farsighted politicians have been thinking of and supporting the concept of hydrogen as an almost ideal chemical fuel, energy carrier, and storage medium.

As a fuel, it doesn't pollute. As an energy-storage medium, it would answer Fuller's call for some method "to store [energy] when it is available for use when it is not available."

Hydrogen is not an energy "source," a mistake still made fairly often by otherwise sophisticated, well-informed people. That is, it is not primary energy (like natural gas or crude oil), existing freely in nature. It is an energy *carrier*—a secondary form of energy that has to be manufactured (like electricity, which doesn't exist freely in usable form either). Hydrogen can be generated from many primary sources—an advantage in itself, since it reduces the chances of creating a hydrogen cartel similar to OPEC (which, for a while at least, was able to dictate global energy prices).

Today, hydrogen is made (that is, extracted) mostly from fossil fuels. But efforts to clean up these fuels (to "decarbonize" them, in the jargon of energy strategists of the 1990s) will increase. To decarbonize really means to adapt techniques long used in the chemical, petroleum, and natural gas industries to strip out the carbon or CO_2 and store ("sequester") it out of harm's way, leaving hydrogen.

In the future, hydrogen will be made from clean water and clean solar energy—and just possibly (though it seems unlikely from the anti-nuclear perspective of the late 1990s) from "cleaner" versions of nuclear energy, including fusion. Since it can be made from both nonrenewable and renewable sources, it can be phased into the overall energy structure by whatever method is most convenient and least wrenching to a given country, state, region, or economy—perhaps, for example, coal gasification in the western United States and solar-based electrolysis in deserts in the Middle East or in the southwestern US. Israeli scientists are testing direct solar water splitting, in which the sun's concentrated heat would break up water molecules into hydrogen and oxygen. Water could be electrolyzed with electricity produced by geothermal resources in some areas, and perhaps also from the oldest form of renewable energy: hydropower.

In the simplest terms, the broad outlines of a future "hydrogen economy" run something like this:

Clean primary energy—probably solar energy in its many variations; possibly an advanced, environmentally more benign version of nuclear energy—would produce electricity to be used to split water into hydrogen as fuel, with oxygen as a valuable by-product. Alternatively, heat produced by solar or nuclear power plants would be used to crack water molecules thermochemically in processes now under development. More exotic methods in which hydrogen is produced from genetically engineered microbes, from algae, and from other biological processes are likely candidates further down the road.

Hydrogen would be used as an energy-storage medium—as a gas under pressure, in hydrogen-absorbing alloys (the above-mentioned hydrides), as a cryogenic liquid, or in activated-carbon materials and carbon nanostructures; but also in the form of relatively conventional fuels, such as methanol. Hydrogen would fulfill the indispensable storage function of smoothing the daily and seasonal fluctuations of solar power.

Hydrogen could be burned in modified internal-combustion engines—jets, turbines, four-strokes, two-strokes, Wankels, diesels. This was the vision, conviction, and message of hydrogen's supporters from the 1970s through the mid 1990s. Since then, with sudden and rapid advances in fuel cell technology, the emphasis has shifted dramatically toward fuel cells as the future engines of choice for transportation[16] and also as clean, efficient, decentralized sources of electricity for buildings. Fuel cells running on reformed[17] gasoline or methanol would produce trace amounts of carbon emissions—much less than internal-combustion engines of comparable power—plus, perhaps, small amounts of nitrogen oxides from fuel processors that generate hydrogen from these carbonaceous fuels. Ultimately, fuel cells operating on pure hydrogen would be quintessentially clean, producing no nitrogen oxides and no hydrocarbons. The only stuff coming out an exhaust pipe would be harmless water vapor (steam), which would immediately return to nature's cycle of fog, clouds, rain, snow, groundwater, rivers, lakes, and oceans. That water could then be split again for more fuel.

As a gas, hydrogen can transport energy over long distances, in pipelines, as cheaply as electricity (under some circumstances, perhaps even more efficiently), driving fuel cells or other power-generating machinery at the consumer end to make electricity and water.

As a chemical fuel, hydrogen can be used in a much wider range of energy applications than electricity. (For example, it is difficult to envision a large commercial airliner powered by electric motors of any conceivable type.) In addition, hydrogen does double duty as a chemical raw material in a myriad uses.

And unlike other chemical fuels, it does not pollute.

Two major goals of international hydrogen research have been to find economical ways of making the fuel and to find out how to store it efficiently onboard a space-constrained car, bus, or truck. During the 1970s and the 1980s, much if not most of the hydrogen-production research was aimed at splitting large volumes of water molecules. This was perceived as the crucial prerequisite to using hydrogen as a fuel. In the 1990s, the emphasis shifted to making hydrogen *energy*—not necessarily ultra-pure hydrogen—an industrial and commercial reality. Thus, much more attention has been paid to improving the steam reforming of natural gas. The efforts of carmakers to use methanol as a sort of hydrogen carrier for fuel cell vehicles are another example. The latter has intrinsic ecological appeal because methanol, today produced industrially from natural gas, can also be made without major impact on the atmosphere ("carbon dioxide-neutral" is the catchphrase) from green plants (biomass) that, in their growth phase, absorb CO_2.[18] A third approach is exemplified by the US Department of Energy's logistics-driven strategy of developing, in cooperation with major carmakers, onboard fuel processors that would extract hydrogen from gasoline and other fossil fuels. The managers of the DoE's "Partnership for a New Generation of Vehicles" (PNGV) argue that this approach would spur a shift toward cleaner energy by using the existing fuel infrastructure as a transitional alternative long before an efficient, widespread hydrogen infrastructure could be put in place.

In past decades, hydrogen advocates believed that a global "hydrogen economy" would begin to take shape near the end of the twentieth century, and that pure hydrogen would be the universal energy carrier by the middle of the twenty-first century. Hydrogen may not completely attain that lofty status in that time frame, but it is certain to play a much larger role—directly as a fuel for fuel cells, indirectly as an increasingly large component of carbon-based fuels such as methanol and other conventional fuels—in the decades ahead. Many see it as an increasingly important complement to electricity; electricity and electrolysis can break water down into hydrogen and oxygen, and hydrogen recombined with oxygen can produce electricity and water again. Each will be used in areas where it serves best—and for a long time to come it will have to compete with, and in fact be dependent on, conventional fossil fuels as its source.

What about nuclear power as a primary energy source for the production of hydrogen? The instinctive short answer from most hydrogen supporters and environmentalists probably is that nuclear power's days have come and gone. As one American anti-nuclear protester (Claire Greensfelder, coordinator of the Berkeley-based group Plutonium Free Future) put it in a CNN interview during the December 1997 Kyoto climate negotiations, "trying to solve climate change with nuclear power is like trying to cure the plague with a dose of cholera." But that wasn't always so. In fact, in the 1970s many in the hydrogen community counted on atomic energy as a source of cheap power with which to split the water molecule. As a cosmic energy dance combining the elementary force that heats the sun and the other stars and the elementary building block of all matter, the concept had an almost mystical elegance. But while a nuclear fire burning far away in the cosmos is one thing, building a nuclear reactor in a populated area is quite another—or so it seemed to the increasingly powerful anti-nuclear forces around the world. In the mid 1970s, orders for new nuclear plants began a sharp decline. And then came Three Mile Island (1979) and Chernobyl (1986). It looked as if those two accidents would be the gravestones of the nuclear age. The debate is not over, though. Some long-term energy thinkers, including some with very good environmental credentials, believe that a second wave of environmentally much more acceptable nuclear power stations may well be inevitable and may become a reality in the twenty-first century.[19]

The 1980s were a bad time for environmentalists and clean energy advocates. In the United States, the Reagan administration was basically apathetic to their long-term planetary concerns, focusing instead on military and geopolitical matters. Interest in clean, renewable energy, including hydrogen, didn't really pick up again until the early 1990s, when worries over environmental issues were mounting. It is probably impossible to give an exact date when that interest got started again, but as good a landmark as any is the publication of Al Gore's book *Earth in the Balance: Ecology and the Human Spirit* (Houghton Mifflin, 1992).

For the international community of hydrogen researchers and supporters, a defining moment came in the spring of 1993, when Japan's government announced its WE-NET (World Energy Network) project, a truly long-range project to help launch hydrogen as the world's clean energy currency of choice. WE-NET was an outgrowth and a redefinition of Project Sunshine, a national multi-dimension alternative energy project begun in 1974. The original announcement said that Project Sunshine was to extend until 2020. It would spend the equivalent of about $2 billion on most aspects of hydrogen energy technology—a level of funding and a truly long-term planning horizon, appropriate to the momentous task of addressing a planetary issue such as global warming, that the governments of Western Europe and North America were neither capable of nor particularly interested in at the time. As it has turned out, however, Japan's annual funding for hydrogen programs so far has been more modest than was expected in the first rush of enthusiasm, both because WE-NET's planners decided to start slowly and modestly, first analyzing what was needed, and because Japan's once seemingly unshakable economy suffered severe setbacks in the ensuing years. Still, WE-NET was, and still is, probably the world's first major hydrogen-centered response by a major industrial country to the growing concerns about global climate deterioration caused by fossil fuels.

Also in the early 1990s, the threat that CO_2 and other trace gases might heat up our planet excessively began to command much more public attention, perhaps (as has already been noted) faster in Europe and elsewhere than in the United States. Since the 1992 Rio de Janeiro Earth Summit (which many regarded as grandstanding but ineffectual), global warming has been reported, discussed, analyzed, dissected, argued, and fought over in countless news stories, interviews, magazines, op-ed pieces, scholarly and popular books, TV programs, and Internet postings.

Whether global warming is a real threat to our world is still somewhat inconclusive. But supporters of renewable, alternative, carbon-neutral, zero-emission energy technologies say it is better to be safe than sorry. In view of the uncertainties about the expected course of the climate, they say it is foolhardy to believe there is no problem at all and to continue with business as usual. Yet the business-as-usual course is the one much preferred and vigorously lobbied for by the world's traditional energy industries and their allies. Ross Gelbspan documented this exhaustively and persuasively in *The Heat Is On* (Addison-Wesley, 1997), outlining the machinations of these industries and their front men to subvert the needed shift to clean energy technologies in order to maintain the profitable status quo.

Greenhouse gases exist in tiny fractions in the atmosphere—only parts per million and even per billion. A minuscule change in concentrations could, it is feared, trigger big, unanticipated, and possibly traumatic change in the atmosphere. The Kyoto supplement of the *New York Times* cited John Firor, an atmospheric scientist at the National Center for Atmospheric Research in Boulder, as comparing the situation to that of a corporation vulnerable to a takeover, where a change in only a few shareholders' votes can mean the difference between the corporation's surviving and being swallowed. "It is a highly leveraged situation," Firor said, nicely adopting Wall Street's jargon to the uncertainties of our collective planetary future. Alan Lloyd, appointed in early 1999 as chairman of the California Air Resources Board and one of the 1990s' pivotal personages on the American hydrogen scene, put it another way. In March 1998, addressing a Society of Automotive Engineers fuel cell workshop in Cambridge, Massachusetts, Lloyd said: "Environmental pollution will likely represent the 'cold war' of the next century."

If hydrogen's benefits as a fuel are so great, the average person might reasonably ask, why didn't hydrogen make significant inroads into our energy systems years or even decades ago? There is no single, simple answer to that question; there is a complex array of interlocking factors. For one, there was no real use for hydrogen as long as there were ample supplies of oil and natural gas and as long as environmental worries were the concerns of a tiny minority. Hydrogen's principal advantage over conventional fuels is that it is emission-free. That, by itself, was not thought to merit a society-wide switch to alternatives of any sort. Fossil fuels were cheap, and hydrogen was as much as several times more expensive. Liquid hydrogen, the coldly exotic stuff that powers the Space Shuttle and experimental BMW sedans today, was a laboratory curiosity four or five decades ago.

Technologically, the level of development was such that producing, handling, and storing hydrogen was complex, difficult, and perhaps beyond the routine abilities of the routine consumer. It still is. Even today, some of the major players in the accelerating "hydrogen sweepstakes," including DaimlerChrysler, argue that hydrogen may have to be made available in some form such as liquid methanol to be user-friendly. The technology is still not perfect, and it is still evolving.

Bringing a technology to maturity takes time. As David Hart of the Imperial College Centre for Energy Policy and Technology in London has observed, "we have only recently become able to operate really well with natural gas." Automobiles have been around for more than 100 years, yet even the best-engineered examples have their occasional glitches and breakdowns. Perhaps most important, societal issues have prevented major progress. For one, replacing an entire technologically advanced energy system with something else is a huge undertaking, spanning decades. It is like trying to change the course of a supertanker with kayak paddles. According to one expert, "the energy system consists of an immense infrastructure, enormous physical and human capital, not only tanks and pipelines and motors, but also people—bankers, auto mechanics, drillers, etc. (and politicians, he might have added), hence it evolves slowly."[20] Phasing in hydrogen would require "innumerable replacements"; substituting fuel cells for internal-combustion engines is only one small aspect.

Perhaps the biggest impediment to change for the better is our value system—what are we willing to pay for. By and large, environmental health is not high on the list. As one American expert with experience in the hallways of Congress and in hands-on alternative energy research (C. E. Thomas, Vice President for Energy and Environment of Directed Technologies, Inc., a consulting group based in Arlington, Virginia) summarizes the issue, hydrogen has not taken off because society does not yet place value on sustainability:

In economic terms, the cost of fuels does not include the externalities of health effects due to urban air pollution, oil spills, ground water contamination, the military cost of defending oil, and, most important, the potential risks of major climate change. Put another way, society has a very high discount rate—we discount any adverse effects that occur in the future.

If the price of coal, oil, and, yes, even natural gas included a full accounting of externalities, then hydrogen would look much more promising overnight. If people had to pay $10/gallon for gasoline or 30 cents/[kilowatt-hour] for electricity to cover fossil fuel damages to our health and environment, then suddenly hydrogen fuel-cell vehicles and hydrogen produced by wind, solar or biomass would look like a bargain. Investors would flock to hydrogen equipment manufacturers. People would convert their SUVs to run on clean-burning hydrogen derived from wind energy at only $2.50/gallon of gasoline equivalent.

A truly sustainable energy future has two attributes: no pollution or greenhouse gas emissions, and no consumption of non-renewable resources. There are only two energy options that meet this sustainability goal: renewable hydrogen and fusion.

Pessimistically, Thomas adds:

Sustainability requires the intervention of governments. Governments alone have the responsibility of protecting the commons. Industry has no major incentive (other than public relations) to build a sustainable energy system. Their overriding objective is return on investment, and burning fossil fuels is very profitable. At best, they will sponsor renewable energy R&D or fuel-cell programs with an infinitesimally small fraction of their profits to give the appearance of preparing for a sustainable future. But most governments do not have the vision or leadership to look into the future and to implement policies that will provide for the welfare of future generations. Certainly, the US federal government is effectively paralyzed, barely able to pass all 13 appropriations bills, let alone tackle any significant societal problem. No US leader has the vision to state the need for sustainability and to follow that vision up with bold implementation programs.

Summarizing, Thomas says:

. . . all the key decisions makers who could influence a transition to clean energy carriers like hydrogen have a very short time horizon: industries have to show a return on investment within a few years, and most elected officials feel that they must show results before the next election—at best six years for a Senator, four years for a President, and only two years for a Representative.

He asks plaintively:

Where do we find the visionary leaders who will look two or three decades into the future and imagine a better world for their children, grandchildren or even great grandchildren?

David Hart is more sanguine. He believes the time is finally at hand when hydrogen will start to make major inroads because of "a confluence of drivers that all point in the same direction—towards hydrogen." The "drivers" include the requirement for a reduction in CO_2 emissions, appalling urban air quality, legislation dictating zero-emission vehicles, progress in fuel cell technology, a move toward the use of local resources for energy production, the need to store intermittent renewable energy, concerns about fossil-fuel resources, and the security of energy supplies. Hart concludes:

There is only one common thread running through these, and that is hydrogen. While other energy carriers can assist in achieving some of these objectives, none of them meet all of the requirements. That is why even the major oil companies are predicting that the energy future is hydrogen.

Equally, the costs are changing. Fossil fuels will not be cheaper to find, and renewables are definitely getting cheaper to use. Storage and use technologies such as fuel cells are already much cheaper, though they have some way to go. However, the advantages seem to be forcing development in the right areas, and the costs of conventional fuels are going up—though not always at the pump. Health and damage costs are much higher than ever before, and people are now starting to consider them, though they may not be added to the price of a gallon of diesel.

Concern that environmental issues may create new and serious global political conflicts does not yet appear to have shown up on Washington's political event horizon or in Washington's parochial politics, but it is discussed among political elites elsewhere. A case in point is that of a young German diplomat, Marcus Bleinroth, a cousin of Frederick Kempe, editor and associate publisher of the European edition of the *Wall Street Journal*. In his recent book *Father/Land* (Putnam, 1999), a highly personal investigation of his German roots, Kempe says of Bleinroth: "As a trained economist, he is convinced the twenty-first century will bring some of the most difficult negotiations ever on ecological matters with developing countries, particularly China. Bleinroth has made himself fluent not only in Chinese but also in environmental issues." Kempe quotes his cousin as follows:

If China continues to grow and develop, global CO_2 emissions will double by the year 2020 because of China alone. That would be catastrophic for the world climate. Ecological problems, however, are finally a question of economic policy. Without an international redistribution of wealth and a greater transfer of resources to developing countries, we have no future. That will require complicated negotiations whose importance and strategic weight might compare with disarmament talks during the Cold War.

Fears about global warming and CO_2 buildup in the atmosphere surfaced decades ago. In 1979, for example, a British Broadcasting Corporation TV documentary about hydrogen energy quoted a meteorologist stationed in

Hawaii (home of one of the world's most important atmospheric CO_2 monitoring posts) as follows: "We don't know at this point whether [CO_2] will build up so that it can do damage. The oil crisis may have slowed it a little. . . . A lot of people believe we could get into trouble, irreversible trouble, in about ten years' time."

Hydrogen contains no carbon at all. Burning it and converting it to energy produces no CO_2 and no greenhouse gas. Used as a fuel, it would reduce and eventually eliminate at least the man-made share of CO_2 deposited in the atmosphere. Switching to hydrogen energy—even perhaps to hydrogen from fossil fuels as a stopgap measure—may help save our children's health and perhaps their lives.

The sky isn't falling—so far. But unless something is done on an international scale, with measures that prove we can actually use our collective human intelligence and wits to guarantee our survival, the time may come when the sky will turn so gray, poisonously yellow, or red from heat and pollution that it might as well be falling. Time will undoubtedly tell.

2

Hydrogen's Discovery: Phlogiston and Inflammable Air

Water is everything. So taught Thales of Miletos (a settlement on the western coast of Asia Minor). Thales, who lived from about 624 B.C. to 545 B.C., was a pre-Socratic Greek philosopher, reputedly the founder of the Milesian school of philosophy. Although he apparently wrote nothing, he was regarded as one of the Seven Wise Men of Greece in his time. The first Western philosopher of record, he is said to have introduced astronomy to ancient Greece. Before Thales, the universe was explained mostly in mythological terms. For Thales, however, water was the primordial material and the essence of everything else in the world. The ideas of Thales, said to be traceable to Babylonian beliefs, are "easily understandable in that the observation of water turning into rigid ice and its transformation into an air-like state led to the thought that all things were derived from matter of middle characteristics."[1] Other early philosophers added air (Anaximenos of Miletos), fire (Heracleitos of Ephesus), and earth (Empedocles of Agrigentum) to the list of elements.

In a way, Thales was not far off the mark. We know now that water consists of two elements: hydrogen and oxygen. Nevertheless, the preponderant part of water is hydrogen (in German, *Wasserstoff*—the stuff of water). Hydrogen is the most abundant material in the universe, the simplest and lightest of the elements. Hydrogen is believed to make up about 75 percent of the mass of the universe and to account for more than 90 percent of its molecules, according to the *New Columbia Encyclopedia*. The Harvard astrophysicist Steven Weinberg says that 70–80 percent of the observable universe consists of hydrogen and the rest mostly of helium.[2]

Hydrogen was first produced, more or less unwittingly, around the end of the fifteenth century, when early European experimenters dissolved metals in acids. However, its classification and description took about 200 years. Many scientists contributed to the unlocking of hydrogen's characteristics, an effort that was closely intertwined with the identification and chemical isolation of oxygen.

Not until the seventeenth century was doubt cast on the notion that air was one of the basic elements. A Dutch physician and naturalist, Herman Boerhaave (1668–1738), was the first to suspect that there is some life-supporting ingredient in the air that is the key to breathing and combustion. "The chemists will find out what it actually is, how it functions, and what it does; it is still in the dark," Boerhaave wrote in 1732. "Happy he who will discover it."[3] In England, the brilliant scientist Robert Boyle (1627–1691) also maintained that "some life-giving substance," probably related to those needed for maintaining a flame, was part of the air. The English physician and naturalist John Mayow (1645–1679) claimed that "nitro-aerial corpuscles"[4] were responsible for combustion.

The realization that both oxygen and hydrogen are gases was long delayed by the phlogiston theory, an early, erroneous attempt to explain the phenomenon of combustion. Promulgated by the German physician and scientist Georg Ernst Stahl (1660–1734) and first published in 1697, the theory held that a substance called phlogiston, which disappeared from any material during the combustion process, imparted burnability to matter. It was believed to be impossible to reduce phlogiston to a pure state. Modern chemistry tells us that to burn a material is to add a substance—oxygen—to it. Stahl held the reverse: that combustion was the release of phlogiston from the burning material. Similarly, he interpreted the reverse chemical reaction (reduction, in which oxygen is removed) as the addition of phlogiston. Even the increase in weight during oxidation, a fairly clear indication that something was added rather than removed, was explained in an altogether artificial fashion: Stahl claimed that phlogiston was so light that it was repelled by the Earth. When phlogiston was removed from a compound, Stahl claimed, the material gained weight because it had lost a component that had lightened it. Stahl, wrote one biographer, "did not hesitate to exclude facts if they violated his ideas: unity of thought was his ultimate goal above all factuality."[5]

Meanwhile, the British preacher Joseph Priestley (1733–1864), the Swedish-German apothecary Carl Wilhelm Scheele (1742–1786), and other scientists had discovered oxygen but had not named the element. Scheele isolated the burnable component of the atmosphere and labeled it "fire air." Sometime between 1771 and 1772, he was the first to produce pure oxygen. It was Scheele's bad fortune that his publisher put off publication of his major work, *Chemical Treatise on Air and Fire*, until 1777. His chief competitors, Priestley and Lavoisier, published their discoveries in 1774. In that year, Priestley discovered oxygen—he called it "dephlogisticated air"—when he heated mercury oxide without the presence of air. The resultant gas produced sparks and a bright flame in a glowing piece of wood kindling. When Priestley inhaled the gas, he "felt so light and well that he regarded it as curative and recommended it as a means of improving the quality of air in a room and as beneficial for lung diseases."

Priestley's and Scheele's experiments came to the attention of France's foremost chemist of the day, Antoine Laurent Lavoisier (1743–1794). Lavoisier, who had been studying gases for years, had noted that during burning both phosphorus and sulfur absorbed part of the surrounding air and gained weight in that process. During a visit to Paris in October 1774, Priestley told Lavoisier about his experiments with mercury oxide. Lavoisier had recently received a letter from Scheele about his discovery of this gas, which makes flames burn "lively" and which "animals can breathe." Lavoisier repeated Priestley's experiments. In 1772, Lavoisier had been among the first to make precise weight measurements to quantify how much "air" disappeared during combustion of phosphorus and sulfur. In an elaborate 12-day experiment, he had heated mercury and air in an airtight retort, producing that same gas that was so conducive to combustion and breathing. Lavoisier labeled this gas "oxygen." He concluded one of his papers by as follows: "We shall call the change of phosphorus into an acid and in general the combination of any burnable body with oxygen as oxidation."

In 1789, Lavoisier, not content to refute Stahl's phlogiston theory with experimental evidence, staged a play in Paris to destroy the theory completely. A German visitor wrote: "I saw the famous M. Lavoisier hold an almost formal auto-da-fé in the Arsenal in which his wife appeared as a high priestess, Stahl as advocatus diaboli to defend phlogiston, and in which poor phlogiston was burned in the end following the accusations by oxygen. Do not consider this a joking invention of mine; everything is true to the letter."[6]

The discovery of hydrogen as an element also proceeded by fits and starts. The Chinese reportedly doubted early on that water was an indivisible element. In the Middle Ages, the famous physician Theophrastus Paracelsus (1493–1541) was apparently the first to produce hydrogen when he dissolved iron in spirit of vitriol. "Air arises and breaks forth like a wind," he is reputed to have said of his discovery, but he failed to note that hydrogen was burnable. Turquet de Mayeme (1573–1655) noted hydrogen's burnability after he mixed sulfuric acid with iron—a phenomenon that was rediscovered by the French chemist and apothecary Nicolas Lemery (1645–1715), who described the burning of the gas as "fulmination violente et éclatente." Still, there was no thought that this gas was an element; rather, it was believed to be some sort of burnable sulfur.

The final isolation and identification of hydrogen was roughly concurrent with the unraveling of the secrets of oxygen in the second half of the eighteenth century, largely because the same scientists were investigating both air and water. Boyle, for instance, was researching artificial gases—"factitious air," as he called them—and was producing hydrogen from diluted sulfuric acid and iron. Boyle did not regard these gases as significantly different from common air; he saw them as a type of air with different characteristics, a view shared by many chemists of those days.[7]

Henry Cavendish (1731–1810), an English nobleman, was the first to discover and describe some of hydrogen's qualities. However, Cavendish did not name the element hydrogen; caught up in the prevailing belief in phlogiston, he thought he had discovered phlogiston in a pure state—a belief he clung to until his death. Taking off from investigations of "factitious air" by other scientists, Cavendish found that there were two different types: "fixed air" (CO_2) and "inflammable air" (hydrogen). Describing these findings in his first scientific paper, which he presented to the Royal Society of London in 1766, Cavendish gave precise readings of specific weight and specific density for both gases. He proved that hydrogen was the same material as "inflammable air," even though it was derived from different metals and different acids, and that it was exceedingly light—about ¼14 as heavy as air.

Hydrogen's buoyancy was quickly put to aeronautical use. "Our colleague has put this knowledge to practical advantage in making navigation in the air safe and easy," said a eulogizing contemporary the year after Cavendish's death.[8] He was referring to Jacques Alexandre César Charles (1746–1823), a French physicist who confirmed Benjamin Franklin's electrical experiments and who became interested in aeronautics. In 1783 Charles flew a hydrogen-filled balloon to an altitude of almost 2 miles. "In fact," said the aforementioned orator, "one can say that without Cavendish's discovery and Charles's application of it, the Montgolfiers' achievement would scarcely have been feasible, so dangerous and cumbersome for the aeronaut was the fire necessary for keeping ordinary air expanded in the montgolfières. . . ."[9]

Cavendish also demonstrated that mixing inflammable air (hydrogen) with air and igniting the mixture with an electric spark produced water and usually a remnant of air. In other experiments, he ignited hydrogen with pure oxygen; when the ratio was right, this yielded only water, thus definitely establishing the makeup of that first "element." Cavendish's experiments involving electric sparks and hydrogen and oxygen, begun in the late 1770s, were not published until the mid 1780s, in his famous treatise *Experiments on Air*.

Lavoisier had been trying for some time to find out the nature of "inflammable air," which he also had obtained by dissolving metals in acid. On combustion of this gas he expected to obtain an acid, but that was not the result. In 1783, Lavoisier heard of Cavendish's work through an intermediary (Charles Blagden, Secretary of the Royal Society). Lavoisier immediately repeated the experiment, but his first attempt failed to impress fellow scientists with its significance. In other efforts, he took the reverse route: splitting water molecules in a heated copper tube. Iron filings in the tube turned black and brittle from the escaping oxygen, and "inflammable air"—a gas that could have come only from the water—emerged from the tube. In a landmark experiment, Lavoisier combined hydrogen and oxygen and produced 45 grams of water. (The water is still preserved in the French Academy of Science.) His classic, definitive experiments proving that hydrogen and oxygen constitute the basic elements of water were done before a large body of scientists in February 1785. In collaboration with other experimenters, he published his major work, *The Method of Chemical Nomenclature*, in which he labeled the "life-sustaining air" oxygen and the "inflammable air" hydrogen.

In 1793, four years after the storming of the Bastille, large-scale economical hydrogen production was invented under the shadow of the uprising and occasioned by the warfare of the competing factions, according to a fascinating historical account presented at the 1986 World Hydrogen Conference in Vienna.[10] Jean Pottier and C. Bailleux (of France's national utilities Gaz de France and Electricité de France, respectively) noted that Guyton de Norveau, a well-known chemist and "representative of the people" of the Comité de Salut Public (Committee for Public Salvation), suggested using hydrogen-filled captive balloons by the army as observation platforms. Norveau, together with Lavoisier, repeated Lavoisier's famous 1783 experiment on a larger scale, prompting the Comité to approve the large-scale manufacture of hydrogen gas. The task was entrusted to another chemist/physicist, Jean Pierre Coutelle. Coutelle built a furnace equipped with a cast iron tube, which he filled with some 50 kilograms of iron filings. Steam was piped in at one end, and hydrogen came out at the other— 170 cubic meters of the gas in the first round-the-clock trial run, which lasted 3 days. Coutelle subsequently set up shop at an army camp at Meudon, close to Paris, where he built a forerunner of what today would be called a hydrogen generator. The first action-ready generator was constructed in early 1794 at Maubeuge. Meanwhile, a collaborator named Conté refined the design into what Pottier and Bailleux called "the army's standard generator." Contemporary drawings mentioned by Pottier and Bailleux depicted a furnace with seven 3-meter-long iron tubes, each 30 centimeters in diameter, containing 200 kilograms of iron cuttings. Water was injected via a seven-way distributor, and the generated hydrogen was washed and cooled with a rotating washer behind which the inventors had installed a dryer-scrubber. The device also included a temperature-control system—75 years before similar systems with similarly sophisticated components were devised for coal gas generators, according to Pottier and Bailleux.

In the early nineteenth century, so-called hydrogen gas was used to light and heat houses, hotels, and apartments, and to supply street lighting. Usually this was not hydrogen at all but essentially carbon-containing gases derived from wood or coal. The confusion was due to the fact that all were lighter than air and were associated with the intrepid balloonists. (Pottier and Bailleux reported that in 1817 there was a "Café du Gaz Hydrogène" across from the Paris town hall, which in fact was lighted by coal gas.)

Lavoisier had been a member of the Ferme-Générale, a financial corporation that leased from the French government the right to collect certain taxes. The system was open to abuse, and some of its members were widely hated by the public. Lavoisier, who was also one of the commissioners in charge of gunpowder production for the government, got caught up in the swirl of charges and countercharges of the French Revolution, and he became one of its victims. In 1794 all the members of the Ferme-Générale were convicted on trumped-up accusations, and Lavoisier went to the guillotine.

3

A History of Hydrogen Energy: The Reverend Cecil, Jules Verne, and the Redoubtable Mr. Erren

On November 27, 1820, the dons of Cambridge University assembled to hear a clergyman's proposal. It is recorded in the transactions of the Cambridge Philosophical Society that Rev. W. Cecil, M.A., Fellow of Magdalen College and of the society, read a lengthy treatise, titled On the Application of Hydrogen Gas to Produce Moving Power in Machinery, describing an engine operated by the "Pressure of the Atmosphere upon a Vacuum Caused by Explosions of Hydrogen Gas and Atmospheric Air." Cecil first dwelt on the disadvantages of water-driven engines (which could be used only "where water is abundant") and steam engines (which were slow in getting underway). The utility of a steam engine was "much diminished by the tedious and laborious preparation which is necessary to bring it into action." Furthermore, "a small steam engine not exceeding the power of one man cannot be brought into action in less than half an hour: and a four-horse steam engine cannot be used [without] two hours preparation." A hydrogen-powered engine would solve these problems, Cecil averred: "The engine in which hydrogen gas is employed to produce moving force was intended to unite two principal advantages of water and steam so as to be capable of acting in any place without the delay and labour of preparation." Rather prophetically, Cecil added: "It may be inferior, in some respects, to many engines at present employed; yet it will not be wholly useless, if, together with its own defects, it should be found to possess advantages also peculiar to itself."

According to Cecil's explanations, the general principle was that hydrogen, when mixed with air and ignited, would produce a large partial vacuum. The air rushing back into the vacuum after the explosion could be harnessed as a moving force "nearly in the same manner as in the common steam engine: the difference consists chiefly in the manner of forming the vacuum. . . . If two and a half measures by bulk of atmospheric air be mixed with one measure of hydrogen, and a flame be applied, the mixed gas will expand into a space rather greater than three times its original bulk."[1]

Cecil went on to discuss the workings of his engine in considerable detail. The *Transactions of the Cambridge Philosophical Society* did not record whether Cecil actually ever built such an engine. In any event, Cecil's proposal was the first known instance of an early technologist's attempting to put the special qualities of hydrogen to work.

Cecil's suggestion came only 20 years after another fundamental discovery: electrolysis (breaking water down into hydrogen and oxygen by passing an electrical current through it). That discovery had been made by two English scientists, William Nicholson and Sir Anthony Carlisle, 6 years after Lavoisier's execution and just a few weeks after the Italian physicist Alessandro Volta built his first electric cell.

In the next 150 years or so, hydrogen's unique properties were discussed with increasing frequency by scientists and by writers of early science fiction. Probably the most famous example, well known in the world's hydrogen community, is Jules Verne's uncannily prescient description in one of his last books of how hydrogen would become the world's chief fuel. *The Mysterious Island* was written in 1874, just about 100 years before research into hydrogen began in earnest. In one remarkable passage, Verne describes the discussions of five Americans during the Civil War—Northerners marooned on a mysterious island 7000 miles from their starting point of Richmond, Virginia, after a storm-tossed escape by balloon from a Confederate camp.[2] The five are the "learned, clear-headed and practical" engineer Cyrus Harding, his servant Neb, the "indomitable, intrepid" reporter Gideon Spillett, a sailor named Pencroft, and young Herbert Brown (an orphan and Pencroft's protégé). The five are discussing the future of the Union, and Spillett raises the specter of what would happen to commerce and industry if the coal supply were to run out:

"Without coal there would be no machinery, and without machinery there would be no railways, no steamers, no manufactories, nothing of that which is indispensable to modern civilization!"

"But what will they find?" asked Pencroft. "Can you guess, captain?"

"Nearly, my friend."

"And what will they burn instead of coal?"

"Water," replied Harding.

"Water!" cried Pencroft, "water as fuel for steamers and engines! Water to heat water!"

"Yes, but water decomposed into its primitive elements," replied Cyrus Harding, "and decomposed doubtless, by electricity, which will then have become a powerful and manageable force, for all great discoveries, by some inexplicable laws, appear to agree and become complete at the same time. Yes, my friends, I believe that water will one day be employed as fuel, that hydrogen and oxygen which constitute it, used singly or together, will furnish an inexhaustible source of heat and light, of an intensity of which coal is not capable. Some day the coalrooms of steamers and the tenders of locomotives will, instead of coal, be stored with these two condensed gases, which will burn in the furnaces with enormous calorific power. There is, therefore, nothing to fear. As long as the earth is inhabited it will supply the wants of its inhabitants, and there will be no want of either light or heat as long as the productions of the vegetable, mineral or animal kingdoms do not fail us. I believe, then, that when the deposits of coal are exhausted we shall heat and warm ourselves with water. Water will be the coal of the future."

"I should like to see that," observed the sailor.

"You were born too soon, Pencroft," returned Neb, who only took part in the discussion with these words.

Of course Verne did not explain what the primary energy source would be to make the electricity needed to decompose water. But in the overall context of nineteenth-century scientific knowledge, Verne's foresight is remarkable.

Hydrogen also figures in a juvenile adventure novel that seems to have been published shortly after 1900 in England. A British scientist interested in hydrogen, W. Hastings Campbell, referred to the book briefly when introducing a hydrogen paper read in March 1933 at Britain's Institute of Fuel. Campbell told his distinguished audience that *The Iron Pirate* by Max Pemberton had made a great impression on him when he was a boy. Pemberton's potboiler described the adventures of a gang of international crooks who owned a battleship that attained terrific speeds due to the use of hydrogen engines—"another instance of the very annoying persistence with which art always seemed to anticipate discoveries," said the account of that meeting in the *Journal of the Institute of Fuel*.

The 1920s and the 1930s witnessed a flowering of interest, especially in Germany and England but also in Canada, in hydrogen as fuel. The evolution of Canada's Electrolyser Corporation Ltd.—today one of the world's leading makers of electrolytic hydrogen plants (it has delivered some 900 systems to 91 countries)—began early in the twentieth century. Around

1905, Alexander T. Stuart, the father of the current chairman, Alexander K. "Sandy" Stuart, began to take an interest in hydrogen energy while studying chemistry and mineralogy at the University of Toronto. Young Stuart and one of his professors, Lash Miller (a former student of the fuel cell's inventor, William Grove), had noted that most of Canada was importing almost all its fuel except for wood. "At the same time, Niagara Falls' hydroelectric generating capacity was being utilized at a capacity factor of only 30–40 percent," Sandy Stuart related in 1996 in the first of a series of lectures bearing his name at the University of Sherbrooke. "The question was, how could such surplus capacity be converted to fuel energy? The obvious answer was electrolysis of water. This led to our first experimental electrolysers."

As it turned out, Stuart electrolysers came into commercial use not to make hydrogen fuel but to make hydrogen and oxygen for the purpose of cutting steel. The first electrolysers were shipped in 1920 to what was then the Stuart Oxygen Company in San Francisco. Four years later, the Canadian government began supporting the use of Stuart electrolysis cells to make fertilizer in British Columbia. From the mid 1920s on, the elder Stuart also developed concepts for the Ontario Hydro utility to integrate hydroelectric energy with coal, coke, or other carbon sources to make "town gas" (carbon monoxide and hydrogen) for domestic heating, to produce a range of synthetic chemicals (including methanol), and to directly reduce iron ore to iron. In 1934 Ontario Hydro built a 400-kilowatt electrolysis plant, and there were plans to heat buildings with hydrogen and even to run test vehicles on it, but that project was terminated after 2 years. All these efforts ended with changes in Ontario's governments, but mostly with Canada's entry into World War II and with the arrival of natural gas on Canada's energy scene after the war.

On the conceptual level, one of the most important figures in those early years was John Burden Sanderson Haldane, a physiologist turned geneticist, longtime editorial board director of the communist newspaper *The Daily Worker*, and in the 1960s an émigré to India and a guru to that country's growing science establishment. In 1923, when he was only in his late twenties, Haldane gave a famous lecture at Cambridge University in which he said that hydrogen—derived from wind power via electrolysis, liquefied, and stored—would be the fuel of the future.[3] In a paper read to the university's "Heretics" society, Haldane said: "Liquid hydrogen is weight for weight the most efficient known method of storing energy, as it gives about three times as much heat per pound as petrol. On the other hand, it is very light, and bulk for bulk has only one-third the efficiency of petrol. This will not, however, detract from its use in aeroplanes where weight is more important than bulk." In the same paper, Haldane prophesied that 400 years in the future Britain's energy needs would be met by "rows of metallic windmills working electric motors which in their turn supply current at a very high voltage to great electric mains." "At suitable distances," he continued, "there will be great power stations where during windy weather the surplus power will be used for the electrolytic decomposition of water into oxygen and hydrogen. These gases will be liquefied and stored in vast vacuum jacketed reservoirs probably sunk into the ground. . . . In times of calm the gases will be recombined in explosion motors working dynamos which produce electrical energy once more, or more probably in oxidation cells."[4] "These huge reservoirs of liquefied gases," Haldane continued, "will enable wind energy to be stored so that it can be expended for industry, transportation, heating, and lighting, as desired. The initial costs will be very considerable but the running expenses less than those of our present system. Among its more obvious advantages will be the fact that energy will be as cheap in one part of the country as another, so that industry will be greatly decentralized; and that no smoke or ash will be produced."

Also in Britain, Harry Ricardo (one of the pioneers in the development of the internal-combustion engine) and A. F. Burstall were among the first to investigate the burn characteristics of hydrogen, and W. Hastings Campbell, the German Rudolf Erren (who spent most of the 1930s in England), and R. O. King (then with the British Air Ministry Laboratory) worked on hydrogen as a fuel.

In Germany, Franz Lawaczeck, Rudolf Erren, Kurt Weil, J. E. Noeggerath, Hermann Honnef, and other engineers and inventors were researching hydrogen and advocating its use as a fuel. Some of these men admitted to being influenced by Jules Verne. Lawaczeck, a turbine designer, became interested in hydrogen as early as 1907. By 1919 he was sketching concepts for hydrogen-powered cars, trains, and engines. Some his inspiration came from contact with his cousin J. E. Noeggerath, an American of German birth who worked in Schenectady, New York, and later in Berlin. Lawaczeck and Noeggerath collaborated in developing an efficient pressurized electrolyzer. In the 1930s, Lawaczeck was apparently the first to suggest that energy could be transported via hydrogen-carrying pipelines. Honnef dreamed of huge steel towers, up to 750 feet in height, each with as many as five 480-foot windmills producing up to 100 megawatts each, which would be stored in the form of hydrogen; however, his concepts were never developed beyond the construction of a 50-meter prototype tower.

In Italy, a 1937 article in the journal *Rivista Aeronautica* mentioned in passing the experimental efforts of the engineer A. Beldimano to adapt liquid hydrogen for use in aircraft engines.

In the United States, Igor Sikorski mentioned hydrogen's potential as an aviation fuel in a 1938 lecture before the American Institution of Electrical Engineers in Schenectady. After predicting the development of a new type of aircraft engine that would permit planes to fly at speeds of 500–600 miles per hour and altitudes of 30,000–50,000 feet, Sikorski said: "If a method of safe and economical production and handling of liquid hydrogen were developed for use as a fuel, this would result in a great change, particularly with respect to long-range aircraft. This would make possible the circumnavigation of the earth along the equator in a non-stop flight without refueling. It would also enable an increase in the performance of nearly every type of aircraft."

One of the earliest and most fascinating efforts involving hydrogen was its use as not only as a buoyancy medium but also a booster fuel for the Zeppelins, the huge German dirigibles that provided leisurely and elegant transatlantic air travel in the 1920s and the 1930s. Normally, these big skyships carried large amounts of liquid fuel (usually a benzol-gasoline mixture) that was used to drive 12- or 16-cylinder engines, which typically propelled a Zeppelin at an altitude of 2400 feet and a speed of not quite 75 miles per hour—provided there was no headwind. Fuel economy was one problem for the Zeppelin; another was how to reduce buoyancy as fuel consumption reduced a ship's weight. According to a 1929 report by the Zeppelin Company, the rule of thumb was that a Zeppelin's captain had to blow off about a cubic meter of hydrogen for every kilogram of fuel burned up during a nonstop cruise, which typically lasted 3–5 days. Better fuel economy could be achieved by certain engine modifications, such as increasing the compression ratios, but the buoyancy problem persisted. The solution was as simple as it was ingenious: Why not burn the blow-off hydrogen as extra fuel together with the main fuel supply? Zeppelin's engineers found that this was feasible. The addition of between 5 percent and 30 percent hydrogen to the main fuel at compression ratios as high as 10:1 produced substantially higher power output—as much as 325 brake horsepower, in comparison with the normal 269 bhp. It also achieved substantial energy savings. The test-bed findings were confirmed by an 82-hour, 6000-mile cruise over the Mediterranean Sea in 1928, during which a fuel reduction of about 14 percent was achieved. Experimenting along the same lines with diesel engines, the Royal Airship Works in Great Britain found that "it was possible to replace almost the whole of the fuel oil by hydrogen without loss of power." On a typical England-to-Egypt trip, an airship would have saved almost 5 tons of fuel oil, according to these experiments. However, neither the British nor the Germans appear to have applied these findings to routine flights to a significant extent.

One of the best-known hydrogen advocates of the 1930s and the 1940s was Rudolf Erren, a brilliant, visionary German engineer who had trucks, buses, submarines, and internal-combustion engines of all kinds running on hydrogen and other fuels, conventional and unconventional. Erren engines were powering vehicles in sizable numbers in Germany and in England. A flinty engineer from Upper Silesia (now part of Poland) with a pronounced disdain for academics and theoreticians, Erren formed his first company, Erren Motoren GmbH Spezialversuchsanstalt, in a grimy industrial section of northern Berlin in 1928. Two years earlier, he had begun to investigate hydrogen and its properties, pursuing an interest that went back to his childhood. When I visited him in Hannover in 1976, he told me that he, like W. Hastings Campbell, had read Pemberton's *Iron Pirate* as a child. As he recalled the book, it "described a pirate group that had kidnapped a German professor who had developed a hydrogen engine which made the pirates' ship much faster than other ships."

Erren had experimented with hydrogen while attending high school in Katowice. His interest in hydrogen carried over as a hobby through his university years in Berlin, in Göttingen, and in England. "During summer vacations when other students went on vacation," he recalled, "I worked in engine workshops to learn something because I wanted to know these things in practice. Theory alone doesn't work." In 1928 he won his first patents, one of them for a hydrogen engine. Erren presented his data at the 1930 World Power Conference in Berlin. According to him, the terms "Erren Engine," "Erren Process," and "Erren System," now largely forgotten, were then officially recognized to differentiate his combustion process from any other.

In 1930, at the invitation of several British firms, Erren went to London to found the Erren Engineering Company. There he continued his work on developing advanced combustion processes that would permit hydrogen to be used alone as a fuel or as a "clean-up" additive to normal fuels. The technique of "Errenizing" any type of internal-combustion process was apparently relatively well known in the 1930s, at least among automotive engineers. Essentially it meant injecting slightly pressurized hydrogen into air or oxygen inside the combustion chamber, rather than feeding the air-fuel mixture via a carburetor into the engine, a method that commonly resulted in violent backfiring. Erren's patented system required special fuel injection and control mechanisms but left the other engine components intact. With hydrogen used as a booster, the Erren system eliminated backfiring and achieved much better combustion of hydrocarbons with higher output and lower specific fuel consumption.

Kurt Weil, a German-born engineer who was Erren's technical director in the 1930s and who in the 1970s was a professor emeritus at the Stevens Institute of Technology, said that the idea of not permitting hydrogen to come into contact with the oxygen of the air before entering the combustion chamber was representative of Erren's "genius." Weil, who had been in the forefront of promoting hydrogen in the 1970s, explained: "When the valves were closed we injected hydrogen, which had a supercharging effect." This engineering approach was revived in the early 1970s.

In the mid 1930s, Erren and Weil proposed to the Nazi government—which by then was concerned with economic self-sufficiency and with reducing Germany's dependence on imported liquid fuels—that most internal-combustion engines be converted to the Erren multi-fuel system. In addition to using carbon-based fuels produced from Germany's plentiful coal by the Fischer-Tropsch and Bergius systems,[5] it would be possible to use hydrogen produced with off-peak power from Germany's closely knit grid of electric power stations, which normally ran at only about 50 percent of capacity. By 1938, when Weil fled Germany and went to the United States, about 100 trucks were running between Berlin and the industrial Ruhr area in the west—a distance of some 350 miles—switching from one fuel to another along the way ("with the truck fully loaded, on a steep incline with a switch in the cockpit," Weil recalled). In an late-1970s interview, Weil recalled that the engine adaptation was not especially difficult, but it was easier for some engine types than for others: "For a six-in-line it was much easier than for a V-type engine." In regard to the conversion costs, he estimated that "under [late 1970s] conditions the cost would have been a few hundred dollars per engine." Erren believed that more than 1000 cars and trucks were converted to his multi-fuel system; however, one British report put the total at between 3000 and 4000.

The German railway system tested a hydrogen-powered self-propelled rail car in suburban operations out of Dresden. The train was powered by a 75-horsepower six-cylinder gasoline engine. It was "much worn" and running harshly, according to a 1932 report by a Reichsbahn maintenance depot; however, when primed with hydrogen the engine developed up to 83 bhp—an increase of 9.7 percent. Powered by pure hydrogen, the engine produced 77 bhp.

Errenization was catching on in England, too. Erren converted Carter-Paterson delivery vans and buses with Beardmore diesel engines to run on hydrogen for better fuel consumption and less pollution. Erren told of an incident involving an Australian commission that spent 2 or 3 weeks in his shops checking his claims and his engines. Eventually, the commission wanted to conduct an open-road speed test with a bus. The site chosen was a hill outside London. According to Erren:

The police there were always on the lookout because the gentlemen from a nearby club drove faster than the thirty miles per hour speed limit. Well, we wanted an official confirmation. . . . The police were pretty well hidden, but we saw them anyway, switched to hydrogen and instead of driving at 30 miles we did 50 or 52 miles up the hill. The police stopped us, told us that they had timed us with a stop watch and we had exceeded the speed limit, which we had to admit. We paid our fine but thanked them profusely, which in turn astonished them until we explained that we now had official proof of our claims.

In 1935, Erren made headlines in the popular British press with news that warmed the hearts of Jules Verne fans. "Secret Fuel to Smash Air Record" headlined the *Sunday Despatch* of March 24, 1935, subheading the one-column story "Non-Stop Round the World with Liquid Hydrogen."

The story reported that engines were being perfected "in secret" in London that would "enable aeroplanes to smash the distance record; make long flights in the stratosphere; and fly non-stop around the world without refuelling." The project never went beyond the concept stage, however. Four decades later, Erren recalled that the prototype plane, a Rolls-Royce-powered De Havilland, was "ready to go," but that the idea fell by the wayside because of disputes as to whether the attempt would be made from Britain or from Germany.

Two other Erren inventions, the "oxy-hydrogen" submarine and the trackless torpedo, attracted some attention in Britain in 1942. The trackless torpedo, fueled by oxygen and hydrogen, was beguilingly simple in concept. Erren started from the realization that torpedoes leave "tracks" of exhaust-gas bubbles. Because hydrogen and oxygen recombined into water vapor, condensing back into the seawater, no bubbles were formed, and thus there was no giveaway trail. And the oxygen-hydrogen-burning submarine eliminated almost entirely the need for big batteries and electric motors for underwater running. Instead, during diesel-powered surface runs the sub's engine also drove an electrolyzer, generating oxygen and hydrogen, which were then stored under pressure. When diving and running submerged, the same diesel engine burned the oxygen and hydrogen without any exhaust bubbles. Weight savings from the elimination of batteries and electric motors translated into the ability to carry more fuel and extended the sub's range—by one report, to as much as 15,000 miles. The generated oxygen was a valuable reserve for the crew in an emergency, and the pressurized hydrogen could be used to blow out tanks for surfacing if other air supplies were exhausted.

Erren was repatriated to Germany in 1945 after the end of World War II. All his personal and business possessions in England had been confiscated during the war. The papers of his company, Deutsche Erren Studien GmbH, in Berlin, had been lost in Allied bombing. After moving to Hannover, where he helped set up a trade association of German plastics manufacturers, he worked for several years as an independent consulting engineer specializing in pollution control, industrial combustion processes, and related areas. None of his engines seem to have survived the war years.[6]

During World War II, interest in hydrogen as a fuel picked up in some parts of the world where fuel supplies were threatened or cut off because of hostilities. In Australia, industrial use of hydrogen was considered early in the war because of wartime demands for fuel oil and because the oil fields in Borneo had been lost to the Japanese. Queensland's government became attracted to hydrogen after the coordinator for public works, J. F. Kemp, learned about hydrogen progress in England and Germany on a 1938 visit to Britain. After Kemp returned, he ordered some studies of his own. However, it was not until the last year of the war that another Australian engineer, J. S. Just, completed a report dealing with hydrogen production via off-peak electricity in Brisbane. The hydrogen was to be used mostly for trucks. The Queensland government authorized construction of an experimental high-pressure plant in Brisbane, but not much was heard about it. The Allied victory in 1945 and the return of cheap oil and gasoline brought hydrogen progress to a halt.[7]

Interest in hydrogen picked up again around 1950 in the context of fuel cells. Francis T. Bacon, a British scientist, developed the first practical hydrogen-air fuel cell (a development that was to be of great significance later in the American space program).

Also in the 1950s, a German physicist was becoming interested in hydrogen as an energy-storage medium. Eduard Justi, a distinguished German electrochemist at the University of Braunschweig, had been working for years on the development of new, more efficient fuel cells. In a 1962 monograph titled *Cold Combustion—Fuel Cells* (Franz Steiner), Eduard Justi and a co-worker, August Winsel, discussed the prospects of splitting water into hydrogen and oxygen, storing these gases separately, and recombining them in fuel cells. Justi later amplified his ideas in the 1965 book *Energieumwandlung in Festkörpern* (Vanderhoeck & Ruprecht), in which he proposed using solar energy to produce hydrogen along the Mediterranean and piping it to Germany and other countries.

In 1962, John Bockris, an Australian electrochemist, proposed a plan to supply US cities with solar-derived energy via hydrogen. Bockris (who in 1975 published *Energy: The Solar-Hydrogen Alternative* (Halstead), the first detailed overview of a future solar-hydrogen economy), says that the term "hydrogen economy"—which has multiple economic and environmental meanings—was coined in 1970 during a discussion at the General Motors Technical Center in Warren, Michigan. Bockris, at the time a consultant to GM, was discussing prospects for other fuels to replace gasoline and thereby help to eliminate pollution, a subject that was then beginning to creep into the public consciousness. The group concluded that "hydrogen would be the fuel for all types of transports," Bockris related in his book. GM did some early experimental work on hydrogen but apparently did not give it the attention—at least, not the degree of publicity—that Daimler-Benz gave it 6 years later.

In 1970 an Italian scientist, Cesare Marchetti, delivered a lecture at Cornell University in which he outlined the case for hydrogen in lay terms. Marchetti, at the time head of the Materials Division of the European Atomic Energy Community's Research Center at Ispra in northern Italy and one of Europe's most persuasive hydrogen advocates, had been calling for the use of hydrogen in Europe and the United States since the late 1960s.[8] Hydrogen, produced from water and heat from a nuclear reactor, could free humanity from dependence on dwindling fossil fuels, Marchetti said at Cornell. "The potential for hydrogen is very great, and a smell of revolution lingers in the air," he told the audience. Marchetti, who has the gift of putting complex relationships into simple terms, stated the hydrogen proposition as follows:

The reason why the studies of industrial utilization of nuclear energy have concentrated on the production of electricity is that a substantial 20 to 25 percent of the energetic input in a developed society is used in the form of electricity and that its production is lumped in large blocks where reactors can show their economies. But almost nothing has been done to penetrate the remaining three quarters of the energy input: food, fuel, ore processing and miscellaneous uses where society is geared to using a wide variety of chemicals.

The problem is to find a flexible intermediate, produced in large blocks in which nuclear heat can be stored as chemical energy and distributed through the usual channels. . . .

In my opinion, the best candidate to perform such a task is hydrogen: on one side hydrogen can be obtained from water, a cheap and plentiful raw material. On the other side, hydrogen can be used directly and very efficiently for:

1. ore reduction, as an alternative to coal,
2. home and industrial heat as an alternative to oil,
3. in chemical synthesis, in particular (for making) ammonia and methanol,
4. producing liquid fuels, such as methanol, for transport; ammonia and hydrogen themselves have potential in the future,
5. producing food, particularly proteins, via yeasts such as hydrogenomonas.

Points one to four cover most of the 80 percent of the energy input, excluding electricity.

Point five can solve once and for all the problem of feeding a growing world population.

A similar message was spread in lectures, papers, and articles in the United States by a few scientists and engineers who had come to the same general conclusion in their respective disciplines. Derek Gregory, a British scientist working at the Chicago-based Institute of Gas Technology, had become interested in hydrogen as a clean substitute for natural gas. Gregory wrote a seminal article on the hydrogen economy for the January 1973 issue of *Scientific American*. He was strongly supported in his work by Henry Linden, founding president of the IGT and éminence grise behind many of the early hydrogen-related R&D efforts in the United States.[9] Bob Witcofski, a young researcher working for the National Aeronautics and Space Administration, had become aware of the exciting prospects of liquid hydrogen as a fuel for aircraft, including nonpolluting supersonic and hypersonic airplanes. Lawrence Jones, a particle physicist at the University of Michigan, had become interested in hydrogen both as an offshoot of his scientific work and because of the rising concern over automotive pollution. Larry Williams, a cryogenic specialist at the Martin-Marietta Corporation, had recognized the usefulness of liquid hydrogen as a fuel. Bill Escher, a former rocket engineer, had come to appreciate hydrogen's potential as a fuel through his involvement with the US space program.

In the early 1970s, while General Motors, Ford, and Chrysler by and large ignored hydrogen's potential as a nonpolluting car fuel (publicly, at least), it captured the attention and the enthusiasm of many American academics, engineers, and automotive enthusiasts. Beginning roughly with the work of the Perris Smogless Automobile Association in California, and with a hydrogen-powered car built by the University of California at Los Angeles that placed second in the 1972 Urban Vehicle Design Competition sponsored by General Motors and other companies, efforts to utilize hydrogen in cars and trucks sprang up in the United States, in Germany, in Japan, in France, and even in the Soviet Union.

The US military was also looking into hydrogen as a fuel. An Air Force program begun in 1943 at Ohio State University eventually culminated in the use of combined liquid hydrogen and liquid oxygen as rocket fuel in the US space program. In 1956, Lockheed began secret work on a long-distance high-altitude reconnaissance plane, a forerunner of the U-2. In a parallel program, the National Advisory Committee for Aeronautics, forerunner of NASA, was gathering actual engine flight data in a B-57 twin-jet bomber operating partially on liquid hydrogen. The Navy had been investigating hydrogen as a fuel for a variety of ships, and hydrogen plus oxygen as a fuel for a deep-diving rescue vessel that would be powered by fuel cells. One revolutionary idea tossed around in the mid 1950s was to use energy from nuclear reactors powering aircraft carriers to make liquid-hydrogen fuel for carrier-based airplanes.

One significant military effort of the 1960s was the Army's Nuclear-Powered Energy Depot, "an early experiment in the hydrogen economy," according to a paper presented in 1974 at a Miami conference dubbed The Hydrogen Economy Miami Energy (THEME). The idea was to develop a portable nuclear reactor that could split water into hydrogen and oxygen in the field, making hydrogen available as a chemical fuel for battle tanks and trucks. It was an outgrowth of the "recognition that the dominant problem in the combat theater is the transportation of petroleum," said John O'Sullivan, then an Army chemical engineer and in the 1990s the manager of a fuel cell program at the Electric Power Research Institute in Palo Alto. The idea was dropped because of efficiency problems and because such a portable nuclear hydrogen plant was a "very vulnerable item" that "needed a lot of people" and lost its main advantage—mobility—if it had to be buried or otherwise protected from attack.

The enthusiasm for hydrogen in the early 1970s was a by-product of growing environmental awareness (especially concern over automotive pollution and the mounting conviction that alternative nonpolluting transportation systems and energy forms were needed) and of the awareness that, with the main sources of petroleum thousands of miles away in the politically volatile Middle East, energy sources closer to home should be looked at.

The oil shock of 1973 announced that the age of cheap, convenient liquid fuel would be coming to an end at some point and that substitutes would have to be found. At first blush, hydrogen seemed to provide an easy, fairly fast answer. Produced via electrolysis "cheaply" from "safe, clean" nuclear reactors (so went the conventional wisdom then), hydrogen could be substituted readily for fossil fuels. Thus, environmental concern and the desire for energy security combined to speed up the investigation of hydrogen.

The reasons for the renewed interest in hydrogen were, of course, different for different people, but the idea of a "totally benign energy metabolism," as Lawrence Jones of the University of Michigan once put it, was certainly a large factor. Hydrogen, Jones observed, had "a kind of gut appeal to people." Jones, a particle physicist, put it more formally in a 1971 article in *Science* magazine, writing that the possibility of using liquid hydrogen as an ultimate replacement for fossil fuels had occurred to him in a casual conversation "related to the logistics and use of large quantities of liquid hydrogen in a cosmic-ray experiment." "In remarking on the drop in price of liquid hydrogen in recent years," Jones recalled, "I noted that the cost per liter was about the same as that of gasoline." As he began reading up on hydrogen in the available literature, he reports, "I recognized that . . . it had an inherent self-consistency and appeal which warranted broader discussion. The conclusion I have reached is that the use of liquid hydrogen as a fuel not only is feasible technically and economically, but also is desirable and may even be inevitable."

In another article, published 2 years later in the *Journal of Environmental Planning and Pollution Control*, Jones said: "It soon became apparent that a surprising number of widely separated individuals and groups had very similar thoughts." That phenomenon broke into the open in 1972 at a spring meeting of the American Chemical Society in Boston, where Cesare Marchetti and a co-worker, Gianfranco De Beni, presented their first thermochemical water-splitting process, and again at the Seventh Intersociety Energy Conversion Engineering Conference in San Diego that autumn.

In its September 22, 1972 issue, *Business Week* ran a multi-page article on international hydrogen developments.[10] (Its effect on the scientists assembled in San Diego was apparently quite riveting. Marchetti later wrote, in a personal note, that "out of 650 participants about 500 were concentrated in the [session] on H$_2$.") *Fortune* carried a longer story 2 months later, and that was followed by articles in *Readers Digest, Time, Popular Science,* and other periodicals.

Hydrogen researchers' recognition that they were not alone reached a climax of sorts on May 6, 1972—the 35th anniversary of the Hindenburg disaster—with the creation of the informal H$_2$indenburg Society, a group dedicated "to the safe utilization of hydrogen as a fuel." Bill Escher, whose name appeared on many of the hydrogen papers published in the United States in the 1970s, was the H$_2$indenburg Society's secretary.[11]

In March 1973, THEME—the first major international hydrogen conference—was held in Miami Beach. At that conference, the groundwork was laid for setting up the International Association for Hydrogen Energy, which has been sponsoring biannual World Hydrogen Energy Conferences ever since. T. Nejat Veziroglu, president of the International Association for Hydrogen Energy, recounted the following in 1994 in his opening remarks at the conference in Cocoa Beach: "In the afternoon of the second day a small group, later to be named 'Hydrogen Romantics,' got together: Cesare Marchetti, John Bockris, Tokio Ohta, Bill Van Vorst, Anibal Martinez, Walter Seifritz, Hussein Abdel-Aal, Bill Escher, the late Kurt Weil, myself and a few other enthusiasts. . . ."[12] After a "passionate, yet deliberate debate," it was agreed "that the Hydrogen Energy System was an idea whose time had arrived." "It was a permanent solution to the depletion of conventional fuels, it was the permanent solution to the global environmental problem," Veziroglu said. "It was Anibal Martinez of Venezuela—incidentally, one who took part in setting up the petroleum cartel OPEC—who urged the founding of a society dedicated to crusade for the establishment of the inevitable and the universal energy system," Veziroglu added. "The rest is history." Officially chartered in the autumn of 1974, the association had about 2000 members by 1979. In 1976 it began publishing a quarterly, which soon turned into a bimonthly and which is now a monthly peer-reviewed scholarly journal, the *International Journal of Hydrogen Energy*.

Governments and international organizations were beginning to take notice. In the United States, where hydrogen research funding did not pass the million-dollar-per-year mark until the mid 1970s, $24 million was budgeted for hydrogen research in fiscal 1978—far too little, in the opinion of hydrogen advocates, who compared it to the $200 million the recently created Department of Energy[13] had allocated that same year for research on how to convert coal into natural gas. Both the disparity and the sentiment have changed little: after declining to slightly more than $1 million in the early 1990s, the DoE's hydrogen program budget had laboriously climbed back up to $24 million level by 1999.

West Germany also began funding hydrogen programs on a small scale, earmarking $2 million between 1978 and 1980. Beginning in 1976, the

Paris-based International Energy Agency began to support hydrogen programs; in 1978, its hydrogen budget stood at somewhat more than $16 million, spread over several years.

Starting in 1972, the European Economic Community spent an estimated 60–70 million Units of Account (roughly $72 million–$84 million) on hydrogen. But with interest in hydrogen cooling again as the oil shock began to fade later in the decade, the European Community cut back, budgeting 13.2 million Units of Account (about $10.2 million) for the period 1975–1979.

Japan's Project Sunshine, a 26-year undertaking begun in 1974, initially seemed to be a truly gigantic alternative energy program, comparable in scope to the Apollo Program. It was supposed to assure Japan of clean and plentiful power by the year 2000. Early reports circulating among Western European scientists in the mid 1970s mentioned total outlays of up to $15 billion over the lifetime of the program, including a hydrogen budget of $3.6 billion.[14]

A perceptible shift in attitude took place among the hydrogen workers as their ranks swelled in the 1970s. At the end of the 1960s and in the early 1970s, a relatively small number of highly idealistic individuals, scattered on different continents, formed a kind of elite international movement. They worked more or less independently, often as an after-hours labor of love. Communication among the members was spotty. Beginning perhaps in the mid 1970s, the scene began to change; more information about hydrogen began to percolate through the international scientific community, corporations, and energy planners as more and more researchers and institutions took up the cause full-time. Because of that growth, institutionalization set in, and with it some disillusionment. In part, the disillusionment had to do with excessively high initial expectations. American researchers and enthusiasts in particular grabbed onto the idea of a hydrogen economy as an almost instant panacea for both energy-security problems and environmental problems. When neither a real hydrogen economy nor the beginnings of hydrogen hardware evolved rapidly within a few years, a letdown was probably inevitable, especially in view of the American penchant for putting new ideas into action immediately. The Europeans were perhaps more realistic (or slower on the uptake, depending on one's point of view). Marchetti and others had argued all along that any fundamentally new energy system required decades—maybe a century—before it could take over a sizable share of the total energy pie.

At the 1976 Miami Beach hydrogen conference, keynote speaker Derek Gregory said that he was more pessimistic about the chances for quick introduction of hydrogen as a universal fuel than he had been 2 years earlier at THEME. NASA's Bob Witcofski, who had also been at THEME, agreed but remained enthusiastic: "Some day, I don't know when, the hydrogen economy is going to come. And I would like, years from now, whenever it is, to be able to stand up and say, all right, our hydrogen airplane has improved 20 percent over the first crude model that was demonstrated back in the year 1990 by a Polack who worked for NASA—and I would like to be that Polack."

As it turned out, neither Witcofski nor anybody else at NASA ever got around to building that "first crude model" during the 1990s. However, the Russians did, in the late 1980s—and so did an octogenarian aviation pioneer in Florida. The Russians did it by adapting one of three engines of a commercial jetliner to liquid hydrogen. William Conrad, a retired Federal Aviation Administration air transport rating examiner (and Pan Am's first director of flight training), did it by converting a little single-engine plane to liquid hydrogen.

In general, the 1980s, at least the first half, were not good for hydrogen— or for alternative, renewable-energy, or environmental issues in general. With the advent of the Reagan administration, the environment and clean energy were relegated to a back burner. The order of the day was beefing up the military (cost be damned), so as to confront the Soviet Union and to fight communism around the globe, and promoting unrestrained growth for the private sector (including fossil-fuel industries) at the expense of larger and ultimately much more important societal issues. Charts based on Congressional Budget Office numbers and prepared in the office of Senator Tom Harkin (D-Iowa), showed that the Reagan administration cut renewable-energy budgets by 80 percent almost immediately after Reagan took office. At the same time, spending on nuclear weapons went, in constant 1992 dollars, from about $5 billion to about $10.5 billion in 1990, peaking at around $12 billion in 1992 (presumably because of budget commitments dating from the Reagan years).

Still, some people never gave up hope and kept up the battle for clean-energy technologies. It is difficult to say exactly when the tide began to turn again, but one landmark event came in the spring of 1986 with the first congressional hearings on hydrogen as fuel in 10 years. The hearings were called by the late Representative George Brown, Jr. (D-California), chairman of the House Science Committee's Subcommittee on Transportation, Aviation, and Materials, to consider the Hydrogen Research and Development Act, a bill introduced earlier by Brown, the principal and perhaps the only hydrogen supporter in the House.[15] Brown launched the bill in part to capitalize on the attention given to the hydrogen-fueled aerospace plane—the ultimately ill-fated "Orient Express"—championed by, of all people, President Ronald Reagan in his State of the Union address earlier in 1986. Brown's bill would have authorized a $200 million, 5-year Department of Energy hydrogen research program, established a technical panel on hydrogen, required better interagency coordination on hydrogen research, and required NASA to set up a comprehensive hydrogen research program. Said Brown: "The major problems in using hydrogen appear to be economic, primarily related to production. New ideas . . . point to potential breakthroughs that may dramatically change the economics of hydrogen production." Predictably, the Reagan administration's witnesses were not impressed. Robert San Martin, the DoE's Deputy Assistant Secretary for Renewable Energy, dismissed the idea, saying "new legislation is not required."

The low point in official interest in hydrogen on part of the US government came in 1987, when the Department of Energy proposed a measly $1 million for hydrogen research and development for fiscal year 1988.[16]

Brown and Senator Spark Matsunaga (D-Hawaii) tried again in May 1987, simultaneously introducing new 5-year, $200 million hydrogen bills and companion bills targeting faster commercialization of fuel cells. Congress did not pass the legislation until 1990 (the landmark Spark M. Matsunaga Hydrogen Research, Development and Demonstration Act, named in honor of the senator after his death).

During the next decade, progress seemed painfully slow to observers of the international hydrogen scene, resembling the kind of forward lurches often ascribed to the dialectical materialism school of history—two steps forward followed by one backward. Still, from the vantage point of the late

1990s, it was obvious that a lot of things had in fact happened during that decade and that progress had been made.

It would take too long to trace in detail all the convoluted steps of international hydrogen activities since the late 1980s, but here are a few highlights, some of which will be discussed in more detail in later chapters.

1987

• A two-year study commissioned by Canada's parliament and by two federal ministries urged Canada to make hydrogen energy technology a "national mission."

1988

• The Soviet Union's Tupolev Design Bureau converted a 164-passenger TU-154 commercial jet partially to liquid hydrogen, operating one of the three engines on the rocket fuel. The 21-minute maiden flight took place April 15.

• In May, in Fort Lauderdale, William Conrad became the first person to fly an airplane (a four-seat Grumman American Cheetah) exclusively fueled by liquid hydrogen.

• The previously unknown high level of hydrogen research activities in the former Soviet Union was highlighted by the first World Hydrogen Energy Conference held in Moscow. Almost half of the conference's 150-odd papers came from Soviet scientists. One seminal paper by two American scientists, Joan Ogden and Robert Williams of Princeton University's Center for Energy and Environmental Studies, predicted that hydrogen produced by means of photovoltaic cells could be economically competitive with coal-based synthetic fuels and even with electricity early in the twenty-first century.

• Ontario's Energy Ministry launched a $600,000 (Canadian dollars) program for first tests of a novel fuel cell, a solid polymer electrolyte fuel cell developed by Ballard Technologies and Dow Canada.

• In Germany, sea trials got underway of a submarine powered by a hydrogen-fueled alkaline fuel cell developed by Siemens.

• The European Community and Canada's Province of Quebec agreed in December on the Euro-Quebec Hydro Hydrogen Pilot Project, a joint study of the feasibility of shipping electrolytically generated hydrogen from Quebec to Europe.

1989

• The National Hydrogen Association was launched in Washington with about a dozen members. By the late 1990s, its membership had grown to more than 60.

• After trying for more than 10 years, two scientists at the Geophysical Laboratory of the Carnegie Institution in Washington announced they had produced hydrogen in a metallic state. Ho-Kwang Mao and Russell Hemsley said they had achieved the feat by compressing the gas under ambient conditions to pressures of more than 2.5 megabars (2.5 million atmospheres).

• An international committee to establish technical standards for hydrogen energy was established in Zurich with the blessings of the International Organization for Standardization.

1990

• The world's first experimental solar-powered hydrogen-production plant became operational at Solar-Wasserstoff-Bayern, a research and testing facility in southern Germany.

• West Germany and the Soviet Union agreed to jointly develop propulsion technology for a liquid-hydrogen-powered prototype jetliner.

• A study by the University of Miami's Clean Energy Research Institute estimated that the environmental costs of burning fossil fuels amounted to about $2.3 trillion for 1990—equivalent to about $460 for every man, woman, and child on the planet. For the United States, the researchers said this was equivalent to adding roughly $1 to each gallon of gasoline in real costs.[17]

• ACHEMA, the world's premier triannual chemical equipment industry show in Frankfurt, added a hydrogen energy section to its exhibits.

• General Motors' Allison Gas Turbine Division began work on a methanol-fueled 10-kilowatt Proton Exchange Membrane (PEM) fuel cell in conjunction with the Los Alamos National Laboratory, the Dow Chemical Company, and Ballard Power Systems, Inc.

1991

• The first test runs of an experimental liquid-hydrogen refueling device for cars and buses began at Solar-Wasserstoff-Bayern's prototype solar hydrogen plant in Bavaria. The goal was to refuel a car in a few minutes.

1992

• A grid-independent solar house that used hydrogen for long-term energy storage in Freiburg, Germany became operational. The house was designed by the Fraunhofer Institute for Solar Energy Systems.

1993

• Ballard Power Systems rolled out the world's first PEM-fuel-cell bus at its facilities in Vancouver.

• Japan unveiled a plan that called for spending about $2 billion over almost three decades to promote hydrogen-based clean energy internationally via its so-called WE-NET (World Energy Network) Project.

• Daimler-Benz and Canada's Ballard Power Systems began a cooperative effort to develop fuel cells for cars and buses.

• The South Coast Air Quality Management District, the air-pollution watchdog agency for the Los Angeles basin, funded a study to look at fuel cell locomotives as a means of reducing pollution.

1994

• Clean Air Now, a nonprofit organization in Los Angeles, won $1.2 million in matching federal funds to set up the first photovoltaics-to-hydrogen demonstration project in North America to produce gaseous hydrogen from solar energy and water to fuel three pickup trucks.

• Daimler-Benz displayed its first NECAR I experimental fuel cell vehicle at a press conference in Ulm, Germany, and New Jersey's H Power Corporation launched its methanol-fueled phosphoric acid fuel cell bus during Earth Day activities in Washington.

• The Saudi-German HYSOLAR demonstration plant, near Riyadh, went into operation.

• The first of four hydrogen-powered buses developed under the Euro-Quebec Hydro-Hydrogen Pilot Project was unveiled in Geel, Belgium.

1995

• Five vehicles running on hydrogen, hydrogen blended with other fuels, or a fuel from which hydrogen was extracted were displayed at the sixth annual US Hydrogen Meeting in Alexandria, Virginia, organized by the National Hydrogen Association.

• The Chicago Transit Authority announced plans for long-term testing of three hydrogen-fueled Ballard PEM-fuel-cell buses.

• Daimler-Benz Aerospace announced plans for an experiment in which one of the two engines of a DO-328 commuter airplane would be converted to run on liquid hydrogen.

1996

• A stationary molten-carbonate fuel cell power plant operating on natural gas and designed by the Energy Research Corporation began generating electricity in Santa Clara, California.

• At an international press conference in Berlin, Daimler-Benz unveiled the prototype of the first consumer-friendly, pollution-free fuel cell passenger car: the NECAR II minivan.

• Toyota unveiled an experimental PEM-fuel-cell version of its popular RAV4 sport-utility vehicle.

• *Element One,* an hour-long American documentary on hydrogen energy, premiered at the World Hydrogen Energy Conference in Stuttgart.

• A Twentieth Century Fox movie about clean abundant hydrogen, *Chain Reaction,* much anticipated in the international hydrogen community, opened to general disappointment.

1997

• At the Detroit Auto Show, the Chrysler Corporation unveiled the mockup of a PEM-fuel-cell passenger car that would be fueled by hydrogen extracted from gasoline or other liquid hydrocarbon fuel by an onboard processor developed by Arthur D. Little, Inc.

• Addison Bain, a widely respected hydrogen supporter and a retired NASA engineer, challenged the belief that hydrogen was the cause of the *Hindenburg* disaster. Bain said that, though hydrogen had obviously contributed to the blaze, the disaster was almost certainly caused by a spark and significant electrostatic activity in the atmosphere at the time, which ignited the impregnated, highly flammable skin of the dirigible.

• Daimler-Benz and Ballard Power Systems announced plans to spend more than $300 million for the joint development and eventual production and marketing of fuel cells for transportation.

• Daimler-Benz unveiled a third-generation PEM-fuel-cell vehicle: the experimental NECAR III, powered by a 50-kilowatt Ballard fuel cell running on methanol.

• Ford Motor Co. joined the Ballard/Daimler-Benz team for the development of commercial fuel cells for cars, trucks and buses.

• The President's Committee of Advisors on Science and Technology, suggesting that hydrogen might become an energy carrier "of importance comparable to electricity" in the twenty-first century, recommended that the US Department of Energy's hydrogen program be expanded substantially.

1998

• General Motors and its German subsidiary, Opel, launched a new global PEM-fuel-cell project for automobiles in Europe. An experimental fuel cell version of Opel's new Zafira minivan was unveiled at the Paris Auto Show.

• Norsk Hydro, Norway's premier utility, chemicals, and metals conglomerate, announced plans to produce electricity by using hydrogen gas as a fuel for steam turbines.

• Iceland's government took steps, in cooperation with Daimler-Benz and Ballard Power Systems, to lay the groundwork for a hydrogen economy.

• The Royal Dutch/Shell Group set up a new International Renewables division and a Hydrogen Team to investigate business opportunities in this area.

• Accompanied by a public show of support from Royal Dutch/Shell, ZEVCO's alkaline-fuel-cell "Millennium Taxi" made its public debut in London.

1999

• Royal Dutch/Shell set up a Hydrogen Division.

• DaimlerChrysler showed a fuel-cell-powered Jeep "concept vehicle." Ford unveiled a fuel cell version of its P2000 research vehicle and a mockup of a fuel-cell-equipped sport-utility vehicle.

• Europe's first hydrogen gas stations were opened in Hamburg and Munich. A similar station was opened at Ford's research labs in Dearborn, Michigan.

• Three multinationals—DaimlerChrysler, Royal Dutch/Shell Group, and Norsk Hydro—teamed up with six Icelandic institutions and companies in setting up the Icelandic Hydrogen and Fuel Cell Company.

• California launched a cooperative Fuel Cell Partnership with carmakers Ford and DaimlerChrysler, fuel cell maker Ballard Power Systems, oil companies ARCO, Shell, and Texaco, and two state agencies, the California Air Resources Board and the California Energy Commission.

2000

• Ballard Power Systems unveiled the world's first production-ready PEM fuel cell for automotive use at the Detroit Auto Show. The 70-kilowatt unit was to be the basic design to be used by carmakers such as DaimlerChrysler and Ford in the first generation of fuel cell cars, promised for 2004. Also at the Detroit Auto Show, Ford showed a fuel cell version of its new small Focus model, and General Motors showed a fuel cell demonstrator version of a new hybrid five-passenger car, the Precept.

• A collaboration of the National Renewable Energy Laboratory and the University of California at Berkeley unveiled an experimental renewable hydrogen-production technique based on the action of green algae.

• At HYFORUM, an international hydrogen energy conference in Munich, BMW executives said that the company planned to offer the world's first commercially available hydrogen-fueled car, a luxury sedan powered by an internal combustion engine, in 2001.

4

Producing Hydrogen from Water, Natural Gas, and Green Plants

For a solar energy project in usually sunny southern California, it wasn't exactly an auspicious beginning. The morning sky over El Segundo was overcast. A light drizzle fell on the 150-plus people—engineers, environmentalists, government types, Xerox corporate executives, curious locals— who had gathered September 26, 1995 on a half-acre site wedged between the buildings of a corporate office park a couple of miles south of Los Angeles International Airport. The occasion was the official dedication of the first operational solar hydrogen facility in the United States, which was making non-polluting gaseous hydrogen fuel for three pickup trucks. The "sunlight-to-tailpipe" Ford Rangers regularly—and legally—cruised the streets of El Segundo and neighboring communities in the Los Angeles basin under experimental permits issued by the California Air Resources Board.

Probably even more disappointing than the weather to James Provenzano, the executive director of Clean Air Now (CAN), the local environmental group that ran the project in partnership with the Xerox Corporation, was the meager media turnout. While the prestigious *Los Angeles Times*, a suburban paper, and a few local radio stations gave the event some coverage, television reporters—on whom CAN had counted to give the project wide exposure—stayed away in droves. "That was the day of the closing arguments of the Simpson trial," Provenzano recalled ruefully later. "The jury went out the next day." Prime-time coverage of the trial of O. J. Simpson, football superstar turned Hollywood celebrity, for the alleged murder of his wife, had preempted just about anything else on local and national TV.

This $2.5 million solar hydrogen project had begun 18 months earlier as a gleam in the eyes of Provenzano, then the manager of environmental programs for Xerox in El Segundo, and Paul Staples, a local environmental activist. Both were working with Clean Air Now, founded and headed by Robert Zweig, a physician from nearby Riverside and a tireless hydrogen advocate for more than 20 years. Zweig, a long-time official of the American Lung Association, had access to data on the ravaging effects of air pollution on humans in the Los Angeles basin and elsewhere. Around the time of the first oil crisis (1973), Zweig had become convinced that one major way to drastically reduce pollution-caused illnesses (emphysema, asthma, lung cancer) and their ballooning societal costs[1] was to substitute a zero-emission fuel such as hydrogen for conventional carbon-based fuels.

Despite the inclement weather and measly media attention, to solar hydrogen supporters the Xerox-CAN project was an auspicious beginning. The ebulliently expansive Staples (who has since then left the project) said: "Hopefully, history will look back at the 1990s as the decade when all this began. I just hope we can claim a footnote in that effort. This project is the first step in implementing the vision of many of you that are here today."

Howard Berman, a Democratic US representative whose district included nearby sections of suburban Los Angeles in the San Fernando Valley, had been trying for a long time to make hydrogen an economic reality. "The Southern California region," he asserted, "is fertile ground for the growth of clean technologies like Clean Air Now's solar hydrogen facility. The projects are ideal for the development of new employment opportunities for those displaced by aerospace layoffs and military base closures."

Berman promoted an idea, first conceived by Robert Zweig, of a "Hydrogen Corridor" in Southern California, of which the CAN-Xerox project would be one of the earliest major components. A map distributed by CAN showed 15 existing, pending, or proposed hydrogen energy-related facilities, including an Allied Signal PEM fuel cell development center in Torrance, Praxair's existing hydrogen-from-natural-gas facility in Ontario, a test bed for renewable energy systems at Edwards Air Force Base, a proposed H Power facility for converting buses to fuel cell power in Van Nuys, and a fuel cell manufacturing facility in San Fernando proposed by Ballard Power Systems.

The CAN-Xerox facility incorporated, in embryonic form, the whole chain of equipment needed to make hydrogen from sun and water to run two pickup trucks for Xerox and one for the City of West Hollywood: Solar radiation, captured by a 48-kilowatt array of fresnel-lens-enhanced photovoltaic system designed by a California firm, the Solar Engineering Applications Corporation, was turned into electrical energy, which was fed into an electrolyzer supplied by Canada's Electrolyser Corporation. The electrolyzer produced about 1500–2000 standard cubic feet (scf) of hydrogen a day from de-ionized tap water in a separate feedwater preparation system. Next, a mist eliminator removed excess water vapor from the hydrogen gas, which was then piped to an interim gas holder (a metal tank) before being conveyed to a compressor, which compressed the gas to 5000 pounds per square inch (psi). After that, the compressed gas was routed to a high-pressure dryer, which removed the last traces of water from the gas before it was fed into a high-pressure storage device holding about 13,000 scf of gaseous hydrogen, also at 5000 psi. Because of the small capacity of the electrolyzer (this was only a demonstration project), but also to meet anticipated future needs for future hydrogen vehicles, additional hydrogen was trucked in periodically by the industrial gas maker Praxair from a production facility in Ontario, about 60 miles inland, replenishing a supplemental 80,000-scf, 2200-psi storage system with hydrogen. Finally, both the supplemental storage system and the high-pressure tank fed fuel into the hydrogen "gas station," which then refueled the supercharged 1993 and 1994 Ford Rangers, which had been modified by Kaiser Engineering of Riverside. The trucks had been outfitted with lightweight pressurized gas tanks, each holding about 2600 scf of hydrogen gas at 3600 psi—equal in energy content to about 5 gallons of gasoline, and good for about 140 miles. In regular service, each truck covered about 100 miles a week, James Provenzano said. The Xerox trucks carried maintenance equipment and tools; the West Hollywood truck was used for environmental audits and to haul recycling bins. The project came to an end in 1997, when Xerox decided it needed the real estate for other purposes. The storage tanks and the liquid-hydrogen pump stayed at the site, permitting the refueling of liquid-hydrogen-powered trucks operated by the company; however, the solar panels, the electrolyzer, and related equipment were moved some 100 miles east to Palm Desert, at the edge of the Mojave Desert, where they became part of a new hydrogen fuel system built by a bus company, for golf carts converted to fuel cell power (golf carts are street legal in some resort towns), and eventually for hydrogen fuel cell buses operated as part of the California Fuel Cell Partnership.

Thousands of miles to the east, in Europe, solar-powered water electrolysis to make hydrogen fuel was also the basic operating principle of two German facilities: Solar-Wasserstoff Bayern (Solar Hydrogen Bavaria) and HYSOLAR (a joint German-Saudi operation with equipment churning out solar hydrogen near Saudi Arabia's capital of Riyadh and in a parallel but smaller research-and-teaching setup on the outskirts of Stuttgart, venue of the 1996 World Hydrogen Energy Conference). These facilities were bigger and more sophisticated than the CAN-Xerox effort.

Solar-Wasserstoff Bayern was backed by industrial heavyweights. It was jointly operated by Bayernwerk (Bavaria's predominant utility), the carmaker BMW, the industrial gas supplier Linde, and the electrical equipment conglomerate Siemens. (Deutsche Aerospace, which had been one of the project's founders, pulled out in the autumn of 1994.)

Launched in 1986, the 12.4-acre SWB facility was nestled in an idyllic valley near the small community of Neunburg vorm Wald, close to Germany's border with the Czech Republic. It was intended as a sort of permanent testing facility for whatever new type of hydrogen technology came off the drawing boards. Powered by arrays of various types of photovoltaic panels rated at close to 350 kilowatts, the complex housed advanced low-pressure water electrolyzers; systems for compressing, purifying, drying, and storing hydrogen and oxygen; gas-fired boilers for testing various mixtures of natural gas and hydrogen; alkaline fuel cells, phosphoric acid fuel cells, and (just before it was closed) new PEM fuel cells; various utility and auxiliary systems; and an experimental liquid-hydrogen filling station for cars that eventually evolved into a fully automatic robotic system.

Apart from the differences in size, scope, and variety of equipment[2] and the fact that SWB's sedans, fork-lift trucks, and other vehicles were restricted to the facility's grounds whereas in California the hydrogen-powered trucks were licensed to operate on public roads, the SWB facility was similar to the CAN facility in its essential purpose: to demonstrate how to make hydrogen from nothing else but sun and water to fuel zero-emission cars. The only element missing was a liquefaction plant. The liquid hydrogen that was used to fuel the experimental BMWs was trucked from a Linde facility elsewhere in Bavaria. The hydrogen generated at the site was used in other experimental applications, such as running fuel cells and boilers and testing hydride systems and other components. The organizers felt that it was too expensive to liquefy the relatively small amounts of hydrogen generated on site by solar power.

Solar-Wasserstoff Bayern's goals were unambiguous: "We are not a production plant," explained Axel Szyska, one of the facility's general managers, in an autumn 1994 interview. "We're a test center, but a test center that deliberately aims beyond lab scale. We're not interested in coffee-cup-size hardware . . . we're intentionally looking for the next step." In the end, those noble intentions were rendered moot by economic reality. Unwilling to continue funding a experimental facility that offered little promise of early economic payoff, and beginning to feel the harsh winds of market deregulation drifting across the Atlantic, Bayernwerk announced in 1998 that SWB would be shut in 2 years.

HYSOLAR, the result of a cooperation agreement signed in 1986 between Saudi Arabia and Germany, was created to investigate solar hydrogen production and utilization. The Saudi team included researchers from the King Abdulaziz City for Science and Technology (KACST) in Riyadh and from three other universities. The Deutsche Forschungsanstalt für Luft-und Raumfahrt (DLR)—the German Aerospace Research Establishment—ran the German side of the project. During the first 8 years, HYSOLAR's total costs were about DM 64 million (about $37 million), split 50-50 between Saudi Arabia and Germany. Total annual spending averaged about DM 8 million ($4.7 million), said Hartmut Steeb, DLR's HYSOLAR manager, in a 1994 interview.

After some 8 years of preparation and testing, some false starts, and some delays, HYSOLAR officially began churning out hydrogen from solar electricity in early 1994. Power was produced by an array of concentrator photovoltaic panels initially rated at 350 kilowatts. (The rating deteriorated over time to about 280 kW, in part because spare parts for one particular photovoltaic subfield ceased to become available, according to Steeb.)

In addition to the main site, about 30 miles northwest of Riyadh, the project included smaller research and test facilities in the Saudi port city of Jeddah and at DLR headquarters in Stuttgart.

In hydrogen production, HYSOLAR achieved an overall electrolysis efficiency of about 69 percent, according to a paper presented at the 1996 World Hydrogen Energy Conference.

HYSOLAR's R&D program covered most aspects of solar hydrogen technology, including catalytic combustion, hydrogen internal-combustion engines, and fuel cell technology. Fundamental research was being done on photoelectrochemistry, electrochemical energy conversion, combustion processes, in addition to hydrogen energy systems.

Over the years, Saudi and German researchers investigated phosphoric acid and alkaline fuel cells and conversion of various internal combustion engines, including a Mercedes-Benz diesel and a small Briggs & Stratton gasoline engine, to direct hydrogen injection, which was regarded as the most appropriate technique to achieve smooth and efficient operation. Unstable combustion phenomena associated with the use of hydrogen as a fuel, with internal and external mixture formation, were studied in a research engine with a transparent cylinder.

Among other topics, HYSOLAR analyzed the potential for decentralized power stations with integrated renewable energy features for remote areas and the economics of hybrid, conventional diesel, and fluctuating renewable energy systems (photovoltaic and wind). HYSOLAR's researchers produced more than 150 publications.

At a Montreal hydrogen conference in the autumn of 1995, Hartmut Steeb announced that the German government had decided to quit funding HYSOLAR at the end of that year. The new official line, dictated in part by pressures to save money to pay for the enormous costs of German unification, was that hydrogen energy was too long-term (50 years or so) and too expensive a proposition. The government saw hydrogen as "the energy for the day after the day after tomorrow," as Steeb put it. But as of late 1999, the Stuttgart and Riyadh facilities were still operating, with funds coming from DLR and KACST. They were focusing increasingly on PEM fuel cells. That year, DLR and KACST signed a memorandum of understanding to continue joint research in this area until June 2001. "HYSOLAR isn't a research ruin, but a very much alive long-term program, even though some of the research priorities have shifted in the context of fuel cells and other energy technologies," said Werner Schnurnberger, a key administrator of the program at DLR.

The Holy Grail: Extracting Hydrogen from Water

Breaking up the water molecule is a difficult, energy-intensive business because of the strong chemical binding forces between hydrogen and oxygen. Hydrogen ordinarily does not exist in a free state; it almost always occurs in compound of some sort. There are many combinations other than the H_2O molecule. Hydrogen gas usually exists only in the molecular state, H_2.

Water can be broken up in many ways, including directly, with extreme heat; with the help of chemicals in two or more steps; with a combination of heat and chemicals; by the action of certain microorganisms found in the oceans; and by running an electrical current through water (or even steam) in electrolysis, a process familiar to high school students everywhere.

Electrolysis is a proven process for making hydrogen and oxygen on an industrial scale. However, it has been used to a significant extent only in places where electricity is very cheap (such as Canada and Norway, with their vast hydropower resources). For a long time, making hydrogen the electrolytic way was considered economically justifiable only when it was to be used as a high-value chemical feedstock, and only when high purity was required. In the late 1960s and the early 1970s, many researchers believed that electrolysis was almost hopelessly outdated and inefficient as a method of making something that would serve, after all, only as fuel to be burned again. But HYSOLAR, SWB, and CAN are clear indications that electrolysis is far from written off—even for making something as mundane as car fuel.

In principle, electrolysis of water is very simple: Two electrodes, one positive and one negative, are immersed in water that has been made more conductive by the addition of an electrolyte, either acidic or basic. When direct-current power is applied, hydrogen begins to bubble up at the negatively charged electrode (the cathode), and oxygen rises out of the solution at the positively charged electrode (the anode). Fresh water of high purity—important because otherwise the accumulating salts would clog up the electrolyzer—is continuously fed to replace the water that has broken down into hydrogen and oxygen.

An industrial electrolysis plant is considerably more complex. In addition to the basic tank arrangement—electrodes, electrolyte, separators to keep the electrodes from touching and shorting out and to separate the two gases at the source, and a container, all of which can take various configurations—such a plant usually requires the following:

• an electric power converter that changes conventional alternating current to direct current at a small but significant loss

• equipment to distribute electric power to the electrodes

• a system of pipes to carry oxygen and hydrogen away from the cells

• a special separation system (in some plants) to remove the gases from the electrolyte

• cooling machinery to remove heat generated during electrolysis

• drying equipment to dry the gases after they have been separated from the electrolyte.

Commercial electrolysis cells date back to the 1890s. However, electrolysis has never been a major industrial-scale activity in the United States, where heavy users of hydrogen, such as producers of fertilizers and the space program, have always relied on cheaper hydrogen made from steam-reformed natural gas. "Electrolysis has traditionally been considered one of the more expensive methods of hydrogen production, and electrolyzers have been assumed to be inefficient and expensive," according to a 1970s survey of hydrogen production and utilization methods by the Institute of Gas Technology.[3] This is not quite the case, though, the report continued. "On the contrary, it is the electric generation step that is expensive and inefficient. Most commercial electrolyzers available today are capable of electricity-to-hydrogen efficiencies above 75 percent, while their capital-cost potential is far less than that of the power stations that would be required to run them." Today many commercial electrolyzers operate at 80–85 percent (of higher heating value; both higher and lower heating value are used to describe efficiencies, leading occasionally to confusion in competing claims), and 90 percent has been demonstrated in the laboratory.

Compared to proposed water-splitting methods such as direct solar water splitting (which requires extremely high temperatures) and thermochemical, electrochemical, or photo-biological hydrogen production techniques (which require sophisticated handling of biological or chemical materials), simple electrolyzers have advantages. Since there are no moving parts, they operate trouble-free and more or less automatically, requiring servicing, such as the exchange of corroded electrodes, only every few years.

Electrolysis also lends itself to operation under higher pressures, which in turn aids process efficiency. Importantly, electrolysis produces very pure hydrogen gas. Fossil-fuel-derived hydrogen is usually somewhat contaminated—potentially problematic for its use in fuel cells. It can be cleaned up, but that adds more steps and therefore increases cost.

The amount of energy needed to decompose water into hydrogen and oxygen by electrolysis is exactly the amount of energy given off in the reverse process, in which hydrogen burns and recombines with oxygen into water vapor. A completely efficient electrolysis cell would require 94 kilowatt-hours to make 1000 cubic feet of gaseous hydrogen. Not all the energy need be supplied as expensive electricity, only 79 kWh. The rest can be brought into the process as simple heat, a less sophisticated and less costly form of energy—an approach that has periodically stirred interest in the idea of electrolyzing steam.

Electrolyzers

There are essentially two types of industrial electrolyzers: the unipolar (tank-type) electrolyzer and the bipolar (filter-press) type. Though there have been improvements in materials, in design, and in conversion efficiency (from about 70–75 percent in the 1970s to about 80–90 percent now, based on the higher heating value of hydrogen[4]), the basic concept has not changed in about 70 years.

An early landmark in electrolysis technology was the decision by a Norwegian utility, Norsk Hydro, to use an electrolyzer for the purpose of making synthetic fertilizer. One of the first truly big electrolyzers, it was built in 1927. The first large electrolyzer in North America was built around 1940 in Trail, British Columbia, by Cominco, Canada's oldest continually operating mining company. After 1945, many large electrolyzers were built around the world. The three biggest (all used in the manufacture of fertilizer) are a plant in Nangal, India, built by the De Nora company of Milan, Italy; the aforementioned Norsk-Hydro plant at Rjakon; and a plant at the Aswan Dam in Egypt, erected by the German company Demag.

The unipolar electrolyzer is the older and simpler type. Alternate electrodes are hung in a tank filled with electrolyte. The electrodes are separated from adjacent opposite-charge electrodes by a diaphragm, usually of asbestos, that allows passage of the electrolyte but prevents mixing of the hydrogen and oxygen gases. The main advantages of the unipolar electrolyzer are these: It requires a relatively small number of parts, the parts are inexpensive, and an individual cell can be shut down for repair or replacement simply by short-circuiting two adjacent cells while the rest of the cells continue making hydrogen. The main disadvantages: It is unsuitable for high-temperature operation because of heat losses due to large surfaces, and it usually requires more floor space than a bipolar electrolyzer.

In a bipolar electrolyzer, each electrode has both a positive and a negative face, with the positive face in one cell and the negative face in the adjacent cell. Proponents say that bipolar electrolyzers take up less floor space than unipolar ones and that they are better suited to high-pressure and high-temperature operation (which is more efficient). Their drawbacks are that they require much more precise tolerances in construction and that they are more difficult to maintain. (If one cell fails, the entire assembly has to be shut down.)

One of the best-known commercial examples of the unipolar cell in North America is the range of so-called Stuart cells, made for decades by the Electrolyser Corporation in Toronto. In the mid 1990s Electrolyser revived the idea of home production of hydrogen for cars. (Daimler-Benz had briefly toyed with the concept in the 1970s in paper studies.) Andrew Stuart, presenting the concept in May 1997 at a Canadian Hydrogen Workshop in Toronto, said the company was developing a washer-dryer-size electrolysis appliance with a built-in compressor that would produce hydrogen overnight. A design was taking shape at the time, and Stuart said the aim was to produce the appliance for about US $1500 (if manufactured in volumes of a million units or more and sold with the car) or for about US $5000 with a production volume of about 10,000. A first prototype of the "Personal Fuel Appliance," under development at the company's new Stuart Energy division, was demonstrated to Ford Motor Company experts in the autumn of 1999, and it was shown to the North American hydrogen community at the spring 2000 National Hydrogen Association conference in Vienna, Virginia.

Teledyne Energy Systems, located near Baltimore, is the preeminent maker of bipolar electrolyzers. Teledyne's electrolyzers range from tiny units for laboratory purposes to big machines capable of generating several tons of hydrogen per day. Teledyne got into the business in the late 1960s when it acquired fuel cell and electrolyzer technology from Allis-Chalmers, which had developed such know-how mainly while working as a contractor for the American space program. Teledyne (bought out by another company in 1993 and renamed Teledyne Brown Engineering) has been credited with being one of the few companies in the United States that continue to invest time, money, and resources into improving standard electrolyzer design.

The year 1995 saw the debut in Germany of an advanced pressurized filter-press electrolyzer designed by the Gesellschaft für Hochleistungs-elektrolyseure, a joint venture of Daimler-Benz Aerospace Airbus, Hamburgische Electricitätswerke, and Norsk Hydro Electrolysers.

Another significant development is the Solid Polymer Electrolyte (SPE) technology,[5] originally spearheaded by General Electric in the 1970s for commercial hydrogen production and for generating oxygen in closed environments such as submarines. Analogous to the proton exchange membrane (PEM) technology used in fuel cells, SPE electrolysis replaces the conventional liquid electrolyte with a solid sheet that looks and feels somewhat like Teflon. When soaked in water, it becomes an excellent conductor. Unlike standard electrolyzers, it doesn't require the addition of acid or alkalis to the water to help drive the electrolysis process. In the early 1980s, when the effects of the 1970s' oil shocks and interest in hydrogen faded, plans for a 5-megawatt demonstration unit fizzled. When the technology languished, GE sold both PEM electrolysis and the PEM fuel cell technology to Hamilton Standard, a subsidiary of United Technologies Corporation. That, in turn, led to the creation of International Fuel Cells as a subsidiary of Hamilton Standard. Over the years, some of GE's core personnel left the company and went elsewhere, dispersing PEM know-how to new players who today are involved in the intense international competition to bring PEM technology to the market.

Recently, PEM electrolysis moved to the fore again with the creation of a new company, Proton Energy Systems of Rocky Hill, Connecticut, to commercialize the technology initially developed by Hamilton Standard. Started in 1996 by a small group of former Hamilton Standard employees, Proton Energy has developed a series of SPE/PEM electrolyzers for industrial use and has also developed fuel cells (including a "regenerative" type that can function reversibly as an electrolyzer and a fuel cell). The Shinko Pantec Company of Kobe, Japan is marketing a similar PEM electrolyzer, dubbed the HHOG (High-purity Hydrogen and Oxygen gas Generator).

The conversion efficiency of PEM electrolysis is about the same as that of conventional electrolyzers with liquid electrolytes: about 80–90 percent. But PEM electrolysis is able to generate hydrogen efficiently and cost-effectively at high pressure without the need for an extra compressor—and with high purity (more than 99.9999 percent). And a PEM electrolyzer is notably compact. Furthermore, PEM electrolyzers, like PEM fuel cells, are claimed to take rapid increases or decreases in electrical input in stride; their proponents say that electrolyzers and fuel cells with liquid electrolytes don't respond very readily to such changes; too much power input too quickly can make the liquid electrolyte bubble, rather than split water into hydrogen and oxygen or generate electricity.

In a further advance, research into high-temperature variants of SPE membrane technology got underway in the mid 1990s at SRI International in Menlo Park, California under contract to Japan's WE-NET program. (Since then, similar research has been begun at other companies and laboratories.) The main advantage, as presented by SRI at a February 1996 workshop in Tokyo, was that new SRI membranes were thermally stable at temperatures of 400–450°C, presumably permitting higher operating temperatures and providing higher yields.

Splitting Steam

Another variant of electrolysis technology was the development of high-temperature steam electrolysis, initially by General Electric and later by Dornier System.

In the late 1960s, General Electric began experimenting with electrolyzing 1000°C (1832°F) steam with an electric current. The main advantage claimed for this type of electrolysis was that at high temperatures the need for expensive, high-quality electricity is drastically reduced because much of the energy would be provided as raw heat. It was hoped that the overall cost of producing hydrogen from water would come down significantly with this approach.

Dornier System began looking into high-temperature steam electrolysis in 1975. The program lasted about 15 years, achieved efficiencies of about 93 percent (according to a press release from the German Science and Technology Ministry), and reduced electricity consumption by about 30–45 percent relative to conventional electrolysis. The core of the system was a solid ceramic oxide electrolyte consisting of zirconium oxide and coated with porous electrodes with a high degree of conductivity for oxygen ions at operating temperatures, with hydrogen evolving on the cathode side and oxygen on the anode side. As the shock waves generated by the oil crises ebbed away and gasoline became cheap again, the program faded into oblivion.

Thermochemical Water Splitting

Thermochemical water splitting—the idea of combining high temperatures with the high chemical reactivity of certain continuously recycled chemicals to break up the water molecule into its components hydrogen and oxygen—was a hot field of hydrogen research in the 1970s, when it first became important to wean the West from its dependence on Middle Eastern oil. It revived a technology that had originated 50 years earlier as an effort to split out hydrogen from water so as to make ammonia for fertilizer. Today the method is of little practical interest, although it is still being investigated in Japan.

One of the earliest proposals for thermochemically producing hydrogen from water was patented in 1924 in England by a scientist working for a company called Synthetic Ammonia and Nitrates Ltd. Emil Collett postulated a two-step cycle. In the first step, water was to react with mercury plus heat to form mercury oxide (HgO) and hydrogen. In the second step, the mercury oxide was supposed to decompose into the original mercury plus oxygen. It probably wouldn't have worked very well; 30 years later, researchers decided that, for a number of theoretical reasons, at least three cycles were needed to break up the water molecule economically and practically by thermochemical means.

In the United States, splitting water in a heat-assisted, closed-cycle chemical process to make hydrogen fuel was first explored in the early 1960s, when an investigation of "chemonuclear" processes was begun at Brookhaven National Laboratory. The idea was to use a large nuclear power plant to produce both hydrogen and ammonia to meet the then-skyrocketing demand for fertilizer. In 1966, building on the idea, two General Motors scientists, James Funk and Robert Reinstrom, wrote a fundamental paper[6] that still crops up in the literature as one of the hydrogen economy's basic documents. Funk and Reinstrom were then working in Indianapolis for GM's Allison Division, which was building tanks for the US Army. The Army was investigating a "Nuclear-Powered Energy Depot" to make hydrogen as tank fuel. The idea, militarily a dud, didn't look very promising scientifically either. "It appears unlikely," Funk and Reinstrom concluded, "that a compound exists, or can be synthesized which will yield a two-step chemical process superior to water electrolysis."

Interest in thermochemistry waned in the 1980s, and since then it has been largely relegated to the sidelines. In the 1990s, General Atomics of San Diego was among the few stalwarts advertising the concept (in the context of applications for a high-temperature nuclear reactor). Work was also still going on at Japan's Atomic Energy Research Institute in collaboration with the University of Tokyo and the Toyo Engineering Corporation.

Producing Hydrogen from Fossil Fuels and by Other Means

At present, the consensus among many of the world's hydrogen proponents is that the initial impetus to really get hydrogen going in the commercial marketplace would have to come from the use of carbon-containing fuels, such as natural gas, biomass, and maybe even coal.

Today about 40 million tons of hydrogen are produced commercially per year around the world.[7] This may sound like a lot, but it really isn't; 40 million tons of hydrogen is roughly equal to 5 quadrillion British thermal units of energy—slightly more than 1 percent of the world's energy demand.

Hydrogen is stripped out of hydrocarbon fuels, typically natural gas, in a method in which natural gas reacts with steam at about 1500–1600°F with the help of a nickel catalyst. The result is a mixture of hydrogen, carbon monoxide, carbon dioxide, steam, and unreacted methane. This mixture is cooled down to about 750°F and reacted further with more steam over a "water gas shift catalyst," producing more hydrogen and converting the carbon monoxide to carbon dioxide. The CO_2 and other impurities are removed by a process called pressure swing adsorption, which leaves pure hydrogen.

A "sorption-enhanced" process being developed by Air Products and Chemicals promises to reduce manufacturing costs by another 25–30 percent by removing CO_2 during the reaction.

A novel method of producing hydrogen from natural gas that surfaced around 1994 is the "Kvaerner process," named after the major Norwegian gas and oil engineering/management company, Kvaerner Engineering s.a., that developed it. (Norway has lots of natural gas as by-product of its offshore oil fields.) The Kvaerner process involves adapting a plasma torch (on which Kvaerner had been working since 1982) in the decomposition of hazardous wastes. The hot plasma would be used to decompose natural gas into hydrogen and commercially saleable carbon black. (Carbon black, used in the rubber, tire, plastics, paint, and ink industries, is also used in some new metallurgical industries, where it serves as a reduction material and a carbon additive or carburizer in steel and foundry operations.) The process is claimed to result in almost 100 percent conversion of natural gas into hydrogen and carbon black. It is said to require relatively little energy and low investment and operating costs, and to be flexible. Kvaerner says it can be adapted to decompose any type of hydrocarbon fuel, from methane to heavy oils. Decomposition takes place at about 1600°C. The reaction can be emission-free: the plasma torch is electricity-driven, and if the primary power is environmentally clean (Norway has a great deal of low-cost hydro power) no emissions will be generated by the conversion process. In a paper presented at the 1996 World Hydrogen Energy Conference, Steinar Lynum, Kvaerner's senior vice president for business development, and Bjorn Gaudernack of Norway's Institute for Energy Technology concluded that, with current prices for natural gas and electricity, "hydrogen from natural gas" was "by far the cheapest option for hydrogen supply." A table said that the cost of producing hydrogen by the Kvaerner process was the equivalent of 6.61 cents per standard cubic meter, versus 6.711 cents for steam reforming. The cost of "sequestering" CO_2 —getting rid of it by burying it deep in exhausted aquifers or in gas or oil reservoirs, or by piping it deep into the oceans—would add 1.793 cent per standard cubic meter to the costs of steam reforming. Generally, the capital costs of the Kvaerner process are not quite double those for steam reforming, according to the above-mentioned table; also, electricity costs would be much higher—0.17 cent for steam reforming, versus 3.3 cents for Kvaerner. However, the economics were claimed to improve with the 5.36 cents/scm credit for carbon black production and sales, plus another 0.5 cent/scm for steam.

Partial oxidation (POX) is conceptually similar to the Kvaerner process, but it uses part of the fossil-fuel feedstock to generate the requisite process heat, yielding only 20–60 percent carbon black; the rest of the combustion products goes up the smokestack as CO_2, nitrogen oxides (NOx), SO_2, and other emissions. Partial oxidation technologies gained ground in the 1990s, often in the context of developing onboard fuel reformers for cars and buses. A partial list of developers of such reformers include Argonne National Laboratory, the Epyx Corporation,[8] International Fuel Cells, Johnson Matthey, Northwest Power Systems, and Shell. A good example of high-temperature partial oxidation of fossil fuels by burning part of the fuel to make hydrogen is provided by Hydrogen Burner Technology, Inc., of Long Beach, California. HBT's president, David Moard, believes that POX technology, though somewhat less efficient than steam reforming, is more manageable and easier to package onboard a car or a truck.

Not only hydrogen but also many other low-polluting alternative fuels, such as methanol and ethanol, could be manufactured from many carbon-bearing sources, including coal and biomass, if the economics were right. Thus, it is quite imaginable that hydrogen could be pushed way back beyond the proverbial back burner if economic interests, for example, were to dictate clinging to the existing production and distribution infrastructure as long as possible. Chris Borroni-Bird, an advanced-propulsion expert at a major American carmaker—an individual who supports conversion to hydrogen in the long run but who understands the realities of the marketplace—has pointed out that old fuel-synthesis methods (such as the Fischer-Tropsch process, which dates back to the 1930s) have supporters even today. The Fischer-Tropsch process was developed in Nazi Germany to make synthetic gasoline and other hydrocarbons from carbon monoxide and hydrogen.[9] "A lot of companies are looking at this and other processes to make gasoline," said Borroni-Bird, who managed DaimlerChrysler's fuel cell hybrid vehicle program before joining General Motors in 2000 as Director of Design-Technology Interface. "Gasoline does not necessarily equate crude oil. As long as oil is cheap—$20 a barrel or thereabouts— these other processes are just not economical, certainly not in the United States. . . . But a lot of companies have these processes or are working on them, ready to be taken off corporate shelves if needed."

In the mid 1970s, in the aftermath of the oil shocks, a number of experts and organizations argued in favor of hydrogen production from coal via coal gasification or coal liquefaction—something the US Department of Energy began looking at again in the 1990s. According to "Hydrogen Tomorrow," a 1976 study by the Jet Propulsion Laboratory: "Hydrogen production from coal is basically a partial oxidation process similar to that for heavy oil. The process is complicated by the necessity to handle a relatively unreactive fuel as a solid and to remove large amounts of ash. The solids-handling problem has a severe impact on costs and prevents much of the technology and equipment developed for petroleum from being used in the conversion of coal. Coal, steam, and oxygen react in the basic gasifier processes to produce hydrogen."

In the United States, General Atomics was the main promoter of the high-temperature reactor (HTR). The company built one HTR, at Fort St. Vrain, Colorado. It had several other orders, but the program was halted in the mid 1970s. In Germany, the Nuclear Research Center at Jülich was the focal point for development of the so-called pebble-bed HTR for both electricity production and coal gasification. A 500-megawatt reactor was expected to go on line by 1990 but never did.

Despite continuing opposition to nuclear energy, General Atomics still has not given up on nuclear energy, both to make electricity and as a source of heat with which to drive chemical processes (including production of hydrogen). At the spring 1996 meeting of the DoE's Hydrogen Technical Advisory Panel (HTAP) in Alexandria, Virginia, General Atomics' project development manager, Malcolm LaBar, asserted that nuclear power was a logical, efficient source of energy with which to make hydrogen. GA's modular helium reactor was meltdown-proof, he said; the fuel particles' ceramic coatings could be safely stored for millions of years without leaking into the biosphere. With a net plant efficiency of more than 47 percent, it could produce electricity for 3.5–3.7 cents per kilowatt-hour, and the reactor's 700–900°C process heat could be used to drive any of the three major hydrogen-production processes (steam reforming of fossil fuels (including methane), high-temperature electrolysis of steam, thermo-chemical water splitting), with efficiencies ranging from 50 percent to 90 percent.

Japan continues to work on producing hydrogen with process heat from a 30-megawatt high-temperature gas-cooled reactor. Papers describing the Japanese effort were presented at the 1996 and 1998 World Hydrogen Energy Conferences and at a 1999 meeting of the Canadian Hydrogen Association.

In the pecking order of ease in making hydrogen from fossil fuels, methane (the main ingredient in natural gas) occupies the top spot because methane usually has been cheap. It is easier to handle industrially than a liquid (such as crude oil) or a solid (such as oil shale or coal). It also contains the most hydrogen: four hydrogen atoms for each carbon atom. Next comes petroleum, with a hydrogen-to-carbon ratio of 1.5–1.6. Oil shale has a ratio of 1.6, but it is solid and therefore more difficult to handle than petroleum. Coal, posing the greatest difficulties for hydrogen extraction, has ratios ranging from 0.72 to 0.92.

The United States' total hydrogen production was 3006 billion standard cubic feet in 1996, according to data compiled by SRI Consulting. Of this, "captive" production accounted for the overwhelming share, including 1211 billion scf for ammonia production, 1120 billion scf for refinery use, 302 billion scf for methanol production, and 117 billion scf for "other" uses. "Merchant production"—production for sale to outside customers (such as NASA and the electronics companies) by industrial gas companies such as Air Products and Chemicals and Praxair—accounted for only 255 billion scf of hydrogen, according to SRI.

According to the 1998 version of SRI's survey of global hydrogen markets, between 1993 and 1997 hydrogen sales in the United States grew at an average annual rate of more than 25 percent. An article titled "Hail hydrogen" in the February 1996 issue of the industry magazine *Chemical Engineering* attributed this growth largely to the demand from oil refineries, which needed more hydrogen to remove sulfur and other contaminants from crude oil. Other big demand drivers, said *Chemical Engineering*, were hydrogen's use in hydrogen peroxide (used to displace chlorine as a bleaching agent in pulp and paper mills) and a host of other applications ranging from waste reduction to purification of synthetic fibers such as Nylon 6 to hydrogenation of specialty chemicals.

SRI's 1998 report said that the upswing in demand for hydrogen was not limited to the United States. Europe also saw an increase in demand, much of it attributable to environmental regulations. Latin America also saw increased demand for the processing of heavy crude oil. In Asia, increases in the overall demand for refinery products pushed sales and production of hydrogen—at least, before the economic crunch of the late 1990s.

The big growth in demand led Air Products and Chemicals and Praxair to construct several very large hydrogen production plants in order to serve refineries and other customers in the United States. Because of the higher production capacity, SRI expects the growth rate of merchant hydrogen sales to drop to about 10 percent per year between 1996 and 2001.

Small-volume users in the metals, electronics, food, glass, and chemical process industries buy hydrogen as a liquid, which is much more economical to transport by truck, and then convert it back to a gas before using it. However, the economics of small-scale hydrogen production and use "may change dramatically over the next five years" with the development of small scale production systems, usually of the partial oxidation (POX) type. "These developments are emerging as a by-product to making fuel cells," the study said, adding, "the costs associated with small-scale hydrogen systems may decline further as fuel cells are adopted commercially, creating demand to support larger production runs of standardized, small scale hydrogen systems."

As the twentieth century drew to a close, the basic consensus was that we will have to rely on making hydrogen from fossil fuels for the next few decades. This certainly is the view of Robert Williams, a senior research scientist at Princeton University's Center for Energy and Environmental Studies, a MacArthur Fellow, and an influential voice in Washington and elsewhere in the ongoing debate about environmental issues and clean energy. In "An evolutionary fuels strategy for fuel cell vehicles," a paper delivered at a 1996 Chicago conference on Commercializing Fuel Cell Vehicles, Williams said that making hydrogen from hydrocarbon fuels is eminently sensible for future large-scale applications, mainly because it would be the least costly production method. In his view, pristinely clean but more expensive electrolytic hydrogen production will play a limited, preferred role in demonstration projects, in niche markets that can be served by hydropower, and in applications where the 99.999 percent purity of hydrogen made from fossil fuels is not good enough. Williams believes that in the foreseeable future—probably well into the twenty-first century—hydrogen and methanol will be produced from petroleum-derived fuels, from natural gas, or from municipal solid waste. Later, hydrogen or methanol derived from biomass and/or coal will enter the market, probably leading to the requirement that CO_2 generated in the process be isolated and sequestered somewhere out of harm's way. "But even with CO_2 sequestration costs taken into account, H_2 derived thermochemically from fossil fuels will be much less costly than electrolytic hydrogen in large-scale markets," Williams foresees. "Hydrogen derived from coal with sequestering of separated CO_2 will be much less costly than electrolytic H_2 when produced on a large scale," and "for the next century it probably would make sense to carry coal to NG (natural gas) fields to produce the H_2, with sequestering of the separated CO_2 in depleted natural gas fields."

In a 1995 paper,[10] Williams and three Princeton colleagues estimated the costs (in 1991 dollars) of producing hydrogen for fuel cell vehicles from natural gas, from biomass, and from coal. Steam reforming of natural gas was the least expensive process, costing $7.17 for a gigajoule (i.e., 10^9 joules) of hydrogen energy. Next came hydrogen produced from gasified biomass, at $8.73/GJ. In third place was a Shell-designed coal gasifier that produced hydrogen at $8.91/GJ. After that came biomass gasifiers by Manufacturing and Technology Conversion International ($8.76/GJ), the Institute of Gas Technology ($10.91/GJ), and Shell ($11.24/GJ). A gallon of gasoline contains 0.131 GJ of energy; Williams et al. said that the comparable price of reformulated gasoline at the pump that year, $1.10/gallon, translated into $8.36/GJ. "The difference between gasoline and hydrogen," they concluded, "was less than a factor of 2, even for the most expensive of these hydrogen production processes. It is the kind of fuel price Europeans and Japanese have been living with—and thriving on—for decades." But that still misses the main point, which is as follows.

Fuel-cell-powered vehicles operated on methanol would be about 2.5 times as efficient as today's internal-combustion engine, Williams et al. said;

with such vehicles operated on hydrogen, the increase in efficiency would approach a factor of 3. Consequently, the bottom line, according to Williams et al., was that biomass-derived methanol and hydrogen would be roughly competitive with today's diesel and gasoline fuels, which are processed from crude oil imported from sources that look increasingly insecure. Biomass-derived methanol and hydrogen also offer zero or near-zero local air pollution and "very low levels of lifecycle CO_2 emissions, if the biomass feedstock is grown sustainably," they say, and "would be roughly competitive" with large-volume coal-derived fuels even if biomass feedstock were relatively expensive. Williams et al. acknowledge that biomass-derived hydrogen and methanol will not be able to compete with natural-gas-based hydrogen and methanol in the near term; however, "natural gas prices are expected to rise substantially over the next decade or so," making biomass nearly competitive.

Finally, according to Williams et al., the production of methanol for export "could provide sustainable new income streams for rural areas in developing regions while bringing competition and fuel price stability to world markets in transport fuels." That idea has been championed by Senator Tom Harkin, hydrogen's leading supporter in the US Senate, who sees the production of fast-growing energy crops such as switch grass as producing a double benefit: a secure domestic supply of energy plus as a new source of income for farmers who depend on federal agricultural subsidies (which are expected to dry up in the coming years in line with Washington's efforts to rein in the US deficit).[11] Thus, the fuel cell vehicle "has the potential to compete with the petroleum-fueled ICEV (internal-combustion engine vehicle) in cost and performance while effectively addressing air quality, energy insecurity, and global warming concerns," Williams et al. conclude.

Not unexpectedly, an earlier study by Robert Williams's Princeton colleague and co-researcher Joan Ogden on the cost of hydrogen produced by the electrolysis of water with electricity derived from renewables, such as solar energy and wind, came up with considerably higher figures per unit of energy produced, based on cost projections of post-2000 photovoltaic, wind, and electrolysis technologies. Nevertheless, the 1993 assessment concluded that there is a crossover point—somewhere around half a million standard cubic feet of hydrogen a day, enough for about 300 fuel cell vehicles—at which electrolytic hydrogen production powered by electricity from photovoltaics or wind would be "roughly competitive" with small-scale steam reforming of natural gas. Produced in these quantities, photovoltaic hydrogen is expected to cost around $12–21 per gigajoule; hydrogen made from water and wind-derived electricity would be higher, about $16–26/GJ. In large-scale production (50–100 million scf per day), hydrogen produced by gasifying biomass would be least costly to produce, at $6–9/GJ—a cost that would be competitive with hydrogen from natural gas (which the Department of Energy projects to cost $4–6/GJ after 2000) and probably less than the cost of hydrogen from coal gasification. Delivering gaseous hydrogen to customers (including compression and storage, pipeline transmission, and local distribution) would add another $6–8/GJ to hydrogen's cost, Ogden says. But "delivered fuel cost alone is not a good indicator of the economic competitiveness of hydrogen as a transportation fuel." Better economic yardsticks are the life-cycle cost of energy services and the "breakeven gasoline price" (the price that would make the life-cycle cost of transportation for a gasoline-powered internal-combustion engine vehicle equal to that of a hydrogen vehicle). "Even if hydrogen is much more expensive than other fuels on an energy basis, it may be able to compete on a lifecycle cost basis for applications where it can be used with higher efficiency than other fuels," Ogden concluded. While "hydrogen fuel cell vehicles would probably cost considerably more than gasoline vehicles," she writes, "our analysis suggests hydrogen fuel cell vehicles might compete on a life-cycle cost basis. Assuming that goals for fuel cells and advanced batteries are achieved, this would occur because fuel cell vehicles would be 2–3 times as energy efficient as gasoline vehicles, would have a longer lifetime and lower maintenance costs."

Many supporters of fuel cells and (especially) of renewable energy are very uncomfortable with the idea of making hydrogen from fossil fuels. They see it as backsliding under the guise of logistical concerns in a Faustian pact that negates the entire reason for going to hydrogen in the first place. Jason Mark, an energy analyst with the Union of Concerned Scientists' transportation program, has questioned whether the infrastructure problem is so severe that "we can't jump it." "Does PNGV really address air quality issues?" he asked at a solar/electric vehicle conference in New York in September 1996. "Probably not. It is sort of secondary as a part of fuel efficiency." Mark's question referred to the US government's much-promoted Partnership for a New Generation of Vehicles with US automakers—a program with the goal of developing a car capable of getting 80 miles per gallon but having no targets in terms of fuel choices or emission reductions. Mark did not see "any appreciable benefits in terms of greenhouse gases from fuel cells until you move to renewable feedstocks." Putting gasoline into fuel cells as a near-term solution could be either "a stepping stone or a stumbling block," he asserted. "Gasoline fuel cell vehicles, for example, could accelerate the penetration of fuel cells in the near term, but by perpetuating our reliance on petroleum (could) merely delay the benefits of a transition to alternatives by maintaining a second-best option," he added. However, others (including Directed Technologies Inc., the Virginia-based consulting firm that was instrumental in persuading the Ford Motor Company to launch a fuel cell program) have shown that hydrogen produced from natural gas would cut greenhouse gases by 40 percent when used in a fuel cell vehicle—much better than with an internal-combustion engine. This means that we don't have to wait for purely renewable hydrogen to start making significant cuts in greenhouse gas emissions.[12]

Beyond the rather conventional hydrogen-production processes discussed above, there are ideas, concepts, experiments, and programs that attempt to produce hydrogen (usually from water) through imaginatively novel approaches, many of them inspired by energy-conversion processes found in nature.

Researchers in the United States, in Japan, and in Europe are investigating processes that use the direct reaction of sunlight with catalysts or biological organisms (such as blue algae), processes that split the water molecule with sunlight and that are aided by a catalyst or semiconductors, the use of "renewable carbon"—the list of plausible and not-so-plausible hydrogen-making processes that have been suggested and/or tried goes on and on. In the mid 1970s, one company (KMS Fusion Inc., of Ann Arbor) even investigated the use of high-powered lasers to initiate thermonuclear fusion and showed that hydrogen could have been produced in that fusion reaction.

One of the first successful photolysis experiments—a milestone in the history of advanced hydrogen-generation processes—was reported in 1972 by two Japanese scientists, Akira Fujishima of Kanagawa University and Kenichi Honda of the University of Tokyo. They managed to produce small amounts of hydrogen by shining light on an electrolysis system that used semiconductor electrodes powered by sunlight only, without any additional electric voltage. In the Fujishima-Honda cell, a titanium dioxide electrode was connected to a platinum black electrode through an external circuit. When the TiO_2 electrode was exposed to light, a current flowed from the platinum electrode back to the titanium dioxide electrode through the outside circuit, and Fujishima and Honda observed gas bubbling up from the electrolyte. From the direction of the current they concluded that water can be decomposed by visible light into oxygen and hydrogen without the application of any external voltage, and that hydrogen was produced at the platinum black electrode and oxygen at the TiO_2 end in minuscule amounts—around 1 percent.

Titanium dioxide is also a key ingredient in an electrochemical photovoltaic system that Michael Graetzel of the Swiss Federal Institute for Technology has been working on since the early 1990s. Dubbed "artificial leaf," the system of large panels and modules mimics natural photosynthesis, but instead of plant matter it produces electricity and hydrogen. In Graetzel's system, TiO_2—an environmentally benign, semi-conductive, low-cost ceramic material widely used in toothpaste and as a white pigment in paints—takes the place of expensive high-purity silicon. In initial tests, Graetzel claimed conversion efficiencies of 7 percent in direct sunlight and 11 percent in diffused light. At the time, Graetzel claimed that the "artificial leaf" may turn out to be the least expensive photovoltaic device yet, producing electricity for about $0.05 per peak watt.[13] In a 1996 report, Graetzel characterized his "chemical solar cell" as a molecular photovoltaic system based on the sensitization of nanocrystalline films by transition-metal charge-transfer sensitizers. The overall efficiency of converting solar energy into electricity was said to be in excess of 10 percent. A similar semiconductor-film-based device split water into hydrogen and oxygen with an overall efficiency of 4.5 percent. Graetzel said that seven industrial organizations were working on commercializing the nanocrystalline solar cell, initially in low-power applications such as a solar watch.[14]

Researchers at the National Renewable Energy Laboratory, in Golden, Colorado, have been experimenting with a hydrogen-evolving photosynthetic bacterium, *Rhodospirillum rubrum*, that produces hydrogen via its hydrogenase enzyme from water even in the presence of oxygen. Normally, hydrogen-producing enzymes in water-splitting microbes are stopped dead in their tracks by oxygen, which inevitably evolves when you break up the water molecule. But in the early 1990s an NREL team led by Paul Weaver discovered a subset of *Rhodospirillum* that continues to pump out hydrogen even when exposed to atmospheric oxygen. But those organisms still don't split water. Coming up with a biological system that will split water entails a set of complex interlocking research tasks, including transferring the genetic properties of *Rhodospirillum* to another organism—a cyanobacterium, for instance. Beyond that, Weaver wants to duplicate the processes that take place in a living cell in a cell-free system "where we take just the light-absorbing, water-splitting complex and link it directly to the hydrogen-producing enzyme," he said in a 1993 interview. In this particular approach, the theoretical efficiency of hydrogen production from water is quite low: about 10–12 percent at maximum. Real efficiencies in real cells are minuscule—only about 0.1–0.3 percent. But, said Weaver, microorganisms are cheap, and "cost is in favor of a biological system. Nature makes the catalysts for doing it very inexpensively from carbon dioxide and water, or from synthesis gas. Operating expenses would be equivalent to those of a farm producing grain crops." Since then, the team's focus has changed somewhat, in part because using cell-free enzymes to produce hydrogen from water is now regarded as impractical and too expensive. In 1998 Weaver said the near-term goal was to use a special group of photosynthetic bacteria that live in darkness to produce hydrogen from carbon monoxide. Taking fuel gases produced from gasified biomass, for example, the bacteria would function as a single-step agent to condition the gas for direct injection into a PEM or a phosphoric acid fuel cell. Most of the energy in those fuel gases consists of hydrogen and carbon monoxide, Weaver explained. The bacteria shift the carbon monoxide into additional H_2, leaving less than 0.1 part per million of residual CO; the resulting hydrogen gas mixture would be clean enough to be fed directly into a fuel cell. In 1999, this technology was moving from the laboratory to an outdoor test area in a first stab at scaling up the process.

Another group of NREL researchers in Golden is pursuing the path first identified by Fujishima and Honda in Japan in the early 1970s. In early

1998, John Turner and Oscar Khasalev reported they had achieved a conversion efficiency of 12.4 percent by splitting water directly with sunlight via a combination of photovoltaic and photoelectro-chemical cells in a single step. In some test runs, "we have seen as much as 15–16 percent, but 12 percent is a good reproducible result," Turner said in his presentation describing the work at the 1998 DoE Hydrogen Program Review. Turner and Khaselev's achievement, first announced in April 1998 in *Science*, made headlines all over the world after the Associated Press picked up the story. *Chemical & Engineering News* quoted chemistry professor Nathan Lewis of the California Institute of Technology as saying that it was a "spectacular number."

In the Turner-Khaselev device, a multilayer cell is immersed in the electrolyte (sulfuric acid diluted with water). When white light shines on the electrode, the top gallium/indium/platinum layer absorbs visible light. A second gallium/arsenide layer underneath absorbs radiation from the near-infrared spectrum, transmitted through the upper layer and a tunnel diode interconnect. All electrons flow toward the illuminated surface, where they break up the water, and hydrogen bubbles off the illuminated surface. Electronic "holes"—vacancies left behind after the electron is excited by light and migrates to the illuminated surface—flow toward the back, with their energy boosted from the gallium/arsenide cell to the point that they can now oxidize water and form oxygen at the platinum electrode. Since hydrogen and oxygen are generated at two physically separated sides of the cell, they can be easily collected and stored. Eventually, Turner hopes to replace the expensive platinum with a nickel or cobalt oxide catalyst being developed by Rick Rocheleau at the Hawaii Natural Energy Institute. Those materials are not as catalytically active, but less costly. Turner estimates that his one-step process is only one-fourth as expensive as other, more conventional two-step solar photovoltaic hydrogen-generation processes, but it is still 3–4 times as expensive as making hydrogen through steam reforming of natural gas. It "isn't the magic bullet that gets us there," said Turner modestly, but it is a "nice scientific accomplishment."

5

Primary Energy: Using Solar and Other Power to Make Hydrogen

"First Large-Scale Solar Energy Project Nears Completion in US Desert" read a headline in the *International Herald Tribune* of November 15, 1977. The story told about progress on an experimental project, Solar One, that eventually was built near Barstow, California. Completed 5 years later, this central receiver pilot plant in the Mojave Desert bundled the combined reflective force of 1818 sun-tracking mirrors, concentrating this solar energy on 24 receivers—solar water boilers—atop a 300-foot tower. These receivers turned water pumped up to them into 960°F steam, which in turn was piped to a turbine generator, producing electricity. Afterward, the steam was condensed back into water and sent back to the receivers. The principal purpose of this installation was to test the concept's basic validity and to test the prototype boiler module, a thermal storage system, and the control system. During the first 2 years of operation, the $144 million, 10-megawatt plant generated 7700 megawatt-hours of electricity at a cost of about $150,000 per month. A retrofit that kept most of the main components in place was completed in early 1998. Renamed Solar Two, the plant continued testing runs for about 2 years until it was shut down in April 1999 because of decreased program funding.

On the other side of the Atlantic, similar but smaller projects were getting underway around the same time. In Sicily, the European Economic Community was completing a 1-megawatt solar electric tower directly connected to the local grid. In Spain, the Paris-based International Energy Agency was building two solar energy plants side by side, one a "solar chimney" plant and the other a "solar farm" plant in which electricity was generated by an array of photovoltaic solar cells; the idea was to compare the efficiencies of the two concepts.

The link between solar power in its various manifestations (wind, wave, and hydroelectric power, photovoltaic energy, and even energy from plant matter) and hydrogen energy is fuzzy in many people's minds. A common mistake, made even by some fairly knowledgeable people, is the belief that hydrogen is a source of energy all by itself. It is not.

To be sure, hydrogen does exist in minuscule percentages in natural gas and therefore could, theoretically at least, be regarded as a "source," but that's useless in practical terms.

Hydrogen is, of course, an energy source in the nuclear sense: In the sun, hydrogen atoms fuse into helium to generate the energy that comes to Earth as solar radiation. But hydrogen is not a *chemical* energy source, even though in the context of a hydrogen economy it is thought of as a chemical fuel—a chemical analogue to electricity as an "energy carrier" with the added advantage that it can be stored over time and thus can help smooth out the fluctuations in the energy supply that are inevitably associated with intermittent primary sources such as the sun and wind.

Electricity is difficult to store in large enough amounts to meet the needs of a city, a region, or an industry. (Here we leave aside for the moment techniques such as "pumped storage"—big reservoirs into which water is pumped when the demand for electricity is low so that it can be used to run turbine generators when the demand is high. Typically, electricity has to be consumed the instant it is generated.)

Another point to remember—one that was pivotal to hydrogen energy advocates during the 1960s and the 1970s—is that electrical energy is still the "minority partner" in the world's energy consumption, accounting for only about a third of the energy used each year. Chemical energy in all its forms—oil, gas, coal, gasoline, firewood, cow patties, synthesis gas— accounts for most of the energy used by the world. This makes it all the more urgent that these forms of chemical energy—each polluting one way or another—be replaced by clean, non-polluting hydrogen.

It is also useful to remember that usable electric power does not exist "raw" in nature but must be manufactured with complicated machinery drawing on "primary sources" (oil, coal, uranium, and, preferably now, solar radiation). "Usable" is the key word here; obviously there is electricity in nature, but it is hard to capture the power of a lightning bolt and make it run some piece of machinery. Similarly, hydrogen must be manufactured in a variety of ways—from natural gas now (the lowest-cost, industrially preferred way), perhaps from coal, but eventually from water and variants of solar energy—photovoltaics, wind, direct solar, solar thermal, and, by extension, biomass.

Loosely speaking, we might compare hydrogen to the transmission in an automobile: It does not generate power by itself, but it makes it convenient to convert the power available from an engine into useful work. Hydrogen works in somewhat the same way but with some extra advantages.

Hydrogen also can also be conceived of as a broad river to which many "primary source" tributaries contribute.[1] At the downstream end, the broad hydrogen river splits again into multiple "irrigation canals" sustaining many economic activities, including transportation, various industries, domestic uses, and chemical activities such as making fertilizer and proteins.

Another useful way of looking at hydrogen may be to think of it as a man-made natural gas minus the global-warming-causing carbon. "Decarbonization," a shorthand term that began cropping up among academic energy strategists in the early 1990s, describes the idea of industrially removing the carbon from fossil fuels before they are combusted and storing ("sequestering") the carbon in form of CO_2 in exhausted gas and oil fields, for example, before it can reach and pollute the atmosphere. "Decarbonized" fossil fuels are, of course, hydrogen. An important implication of this elegant and simple but difficult-to-implement idea is that types of primary sources can be utilized for end uses that would not be possible without the hydrogen intermediary. With hydrogen, nuclear energy, solar energy, or hydropower can be converted to chemical energy to propel cars and airplanes, for example. Without hydrogen, these primary sources are good only for making electricity; it is hard to imagine a big passenger plane powered by electric motors.

Storability is another important attribute of the chemical fuel hydrogen. Hydrogen makes it possible to economically store over time—for the winter season, for example—energy derived from intermittent sources such as solar power. Hydrocarbons (natural gas, petroleum, coal) obviously are easy to store. But how do you store sunlight or the heat from a nuclear reactor? Storage works very well in solar power tower plants, where heat is stored very efficiently in 24-hour, day-and-night cycles in molten salt storage tanks.[2]

But it doesn't work very well in "retail" applications, such as automobiles and trucks, or in long-term seasonal storage. On the other hand, electricity can be stored indirectly if it is used to split water into hydrogen and oxygen. Hydrogen and oxygen can be stored easily, like other industrial gases. By burning the hydrogen later with the stored oxygen, or with oxygen from the atmosphere, electricity or heat or both can be produced again, when and where it is needed, by means of various types of combustion machines (internal-combustion machines, like a car engine, or external-combustion machines, like a steam engine)—or, more efficiently, by recombining the two gases electrochemically, silently and cleanly, without any open flame, in fuel cells.

As far back as the 1970s, the eminent electrochemist John Bockris,[3] founder of the Center for Electrochemistry and Hydrogen Research at Texas A&M University and in earlier years one of most active international proponents of a hydrogen economy put the relationship between primary sources and hydrogen as follows:

The likely sources of energy for the future are atomic and solar. Atomic reactors can provide electricity which would be cheaper as the reactors increase in size, but with size comes the difficulty of thermal pollution, so that large atomic reactors, which would give relatively cheap electricity at source, would have to be placed either on the ocean, far from population centers, or in remote areas such as Northern Canada, Siberia, or Central Australia.

Correspondingly, massive solar collectors are likely to be far from the population centers which need them, for they would be most advantageously situated in North Africa, Saudi Arabia, and Australia. Hence, the electricity to which they would give rise is liable to have to travel at least 1000 miles (1,600 kilometers), and, in some situations, as much as 4000 miles (6400 kilometers), to go from the site of production to the site of use.

The likelihood of this situation, and the energy loss in conduction, gave rise to the concept of a "Hydrogen Economy." Thus, it could be cheaper to convert electrical energy, which will be a product of solar and atomic reactors, to hydrogen at the energy source. Thereafter, the hydrogen would be transmitted through pipes—the pumping energy being relatively small—and converted back to electricity at the site of use (fuel cells) or used in combustion to provide mechanical power.[4]

Since the mid 1980s, nuclear power has lost almost all its luster as a source of primary energy for electricity production.[5] Aside from deep fears among large segments of the population and misgivings about storing highly toxic nuclear waste over tens of thousands of years, nuclear power has turned out to be a huge economic millstone around the collective necks of utility managers. Instead of "too cheap to meter" (the rallying cry of early nuclear enthusiasts), nuclear power has turned out to be "too costly to manage," Paul Gunter of the Washington-based Nuclear Information and Resource Service wrote in a compilation of articles by renewable energy specialists.[6] Gunter quoted a February 1985 article in *Forbes* magazine, "Nuclear Follies," as saying: "The failure of the US nuclear power program ranks as the largest managerial disaster in business history, a disaster on a monumental scale. The utility industry has already invested $215 billion in nuclear power, with an additional $140 billion to come before the decade is out, and only the blind, or the biased, can now think that most of the money has been well spent. It is a defeat for the US consumer and for the competitiveness of US industry, for the utilities that undertook the program and for the private enterprise system that made it possible."

But then there is nuclear fusion—a long-shot, dark-horse option. Unlike nuclear fission, fusion produces no radioactive fuel waste, although the reactor itself would become radioactive over time. Also unlike fission, fusion produces no bomb-grade materials, has no chance of inflating into a "runaway" critical reaction and has a virtually unlimited fuel supply. Fusion reactors would literally run on seawater: The primary reaction requires deuterium and tritium, the two heavy isotopes of hydrogen. Deuterium is extracted from seawater, and tritium is made in a reactor from lithium, also found in seawater. Fusion would produce no local air emissions and no greenhouse gases, if scientists and engineers ever manage to develop an economic method of controlling the thermonuclear reaction. Hydrogen could be generated in a fusion reactor itself, providing another future pathway toward a sustainable energy future with no environmental damage and virtually unlimited fuel supply.[7]

For the moment, however, solar and renewables represent the best complementary energy resource for hydrogen production in terms of environmental benefits.

Some of the more progressive fossil fuel companies are beginning to see the light. John Browne, chief executive officer of British Petroleum, the world's third-largest oil company, was one of the first among his peers to sign on as a believer in the reality of CO_2-caused global warming; he is setting his company on a renewable-energy course, according to *Business Week* ("When green begets green," November 10, 1997). Browne believes that renewables could meet 5 percent of the world's energy demands in 20 years and 50 percent by 2050, and that BP should get in on the ground floor. BP already has spent $160 million on developing its solar energy business and has 10 percent of the world's market for solar power. "The more we thought about this, the more we found that we could improve our environmental performance . . . and actually get higher financial returns," Browne told *Business Week*. The Royal Dutch/Shell Group broke ranks with fossil-fuel orthodoxy too. In quick succession, it set up a new Renewables division (funded at $500 million) in the autumn of 1997, followed in the spring of 1998 by a decision to quit the exquisitely misnamed Global Climate Coalition (an association of oil and other companies dedicated to the proposition that global warming has no relation to growing fossil-fueled CO_2 emissions). Shortly after that, Shell began to take a serious look at hydrogen as an energy carrier. In June it dispatched a platoon of executives to the 1998 World Hydrogen Energy Conference in Buenos Aires for a behind-the-scenes briefing session (an exercise apparently launched at the request of Shell's chairman, Mark Moody-Stuart, who took office on July 1).

In principle, solar energy can be employed to make hydrogen in a variety of ways. Typically, electricity is produced first; it then can be used in electrolysis. It is not clear whether there is one best solar technology for making hydrogen, and many variables come into play— including capital costs, the commercialization status of a given technology, local circumstances (e.g., special tax breaks), the total chain of conversion efficiencies from the amount of sunlight available in a given location, average wind speeds, the growth rate of plants (for biomass production), the conversion efficiencies of various types of photovoltaic cells, the efficiencies of water splitting and hydrogen storage, and the price and usefulness of the by-product oxygen.

Probably the most detailed, most exhaustive body of information relating the various primary sources to hydrogen production is to be found in a series of papers and studies on hydrogen energy systems published since 1989 by the Center for Energy and Environmental Studies at Princeton University, many of them under contract to the US Department of Energy. Many, perhaps most, of those reports listed Joan Ogden as principal author or co-author. Ogden started as a nuclear fusion physicist; however, after recognizing the difficulty of the scientific and engineering challenges facing fusion she decided that she wanted to better understand energy and environment problems generally and "all the options for a clean energy future." She abandoned fusion and began working on hydrogen in 1985. Since then, she has become arguably the world's premier systems analyst for hydrogen energy.

One of Ogden's papers, presented at the January 1997 World Car Conference in Riverside, California, examined the technical feasibility and the economics of a reasonably realistic hydrogen refueling infrastructure for at least a portion of the zero-emission vehicles that will soon be mandated in Southern California. Ogden's estimates of energy requirements were predicated on the higher efficiencies of fuel cells, not on the lower efficiencies of conventional internal-combustion cars running on hydrogen. For starters, she says, potentially ample sources of hydrogen are already in place in the Los Angeles Basin: hydrogen plants and refineries.

By 2010, the South Coast Air Basin is predicted to be home to some 9 million cars (11 million if you count sport utility vehicles), of which more than 700,000 will have to qualify as Zero-Emission Vehicles. If one assumes an "aggressive commercialization scenario" for fuel cell vehicles beginning around 2003, and if one assumes that half of the "zero-emission" cars and 10 percent of all new buses would be powered by fuel cells, this translates into approximately 350,000 fuel cell cars and 300 buses cruising around the Los Angeles area by 2010, requiring about 55 million standard cubic feet of hydrogen per day—about the amount of hydrogen now produced per day at a good-sized oil refinery.[8]

Several hydrogen plants in the area are producing large quantities of hydrogen. Most of their output is already committed to customers, but there is some extra capacity. Hydrogen produced with that extra capacity could be sold as transportation fuel. Also, a number of oil refineries in the area are producing huge amounts of hydrogen, most of which is used internally in refining but several million cubic feet of which is typically available for off-site sale. With new hydrogen facilities planned to meet the expected demand for more reformulated gasoline, extra reformer capacity is being constructed now. Thus, says Ogden, hydrogen from existing sources "could be significant in getting hydrogen vehicles started." The amount available in the region could range from 5 million to 15 million standard cubic feet per day—"enough for 30,000 to 100,000 fuel cell cars or 700 to 2000 fuel cell buses." Depending on what methods are used in capitalization accounting, the cost could be as low as $1 per 1000 scf ($2.8 per gigajoule) at the hydrogen source and as high as about $3 per 1000 scf if certain key capital costs are included. The delivered price at a refueling station could range from a high of slightly less than $4 per GJ (a conventional reformer making hydrogen on site from natural gas) to a low of slightly more than $1 per GJ (same technology, bigger throughput and capacity). On the other hand, liquid hydrogen delivered by truck could cost as much as $30 per GJ—probably tolerable only for the first demonstration projects, without the capital expenses of an on-site natural gas reformer or electrolysis plant. Under certain conditions, a local pipeline could deliver fairly low-priced hydrogen, ranging from $3/GJ (refinery excess) to $8–10/GJ (hydrogen from biomass, from coal conversion, or from municipal solid waste). One dollar per gallon of gasoline translates into approximately $7.67 per gigajoule. In other words, the various prices for hydrogen (except for the expensive liquid variety) could range from lower than to roughly 3 times gasoline's perversely low 1998 benchmark price of $1 per gallon.

Additionally, there are "ample natural gas resources" in the Los Angeles area, writes Ogden. The 55 million scf/day needed for the initial 350,000 PEM-fuel-cell cars could be produced by steam reforming about 22 million scf/day of natural gas—"less than 1 percent of the total natural gas flow through Southern California Gas's distribution system."

And in Southern California there is a "large potential" for using off-peak power (power generated between 6 P.M. and 10 A.M.) to electrolyze water. Ogden cites estimates by Southern California Edison that 4000–6000 megawatts might be available every night, at prices ranging from 4 cents per kilowatt-hour for small commercial customers to 3 cents per kWh for large ones. Translated into electrolytic hydrogen, that amount of electricity would be enough to fuel from 3.5 million to 5.3 million fuel cell cars.

In an earlier paper on the same subject, Joan Ogden, Adam Cox, and Jason White took a closer look at the hydrogen requirements for the Los Angeles Basin after 2010, during a transition to renewable hydrogen. In general, Ogden et al. said, the supplies of natural gas would probably be sufficient for up to several million PEM-fuel-cell cars for several decades. Renewables would have to be phased in if the entire fleet of 12–14 million cars expected for the area were required to be zero-emitters. Electrolytic hydrogen could be produced by means of solar electricity, meaning mostly photovoltaics. In the Los Angeles area, the authors estimate that about 21 square miles of land would be needed to produced about 159 million scf/day of hydrogen, enough for a million fuel cell cars, assuming average annual insulation of 220 watts per square meter, 15 percent photovoltaic efficiency, and 80 percent electrolyzer efficiency. With expected improvements in the cost of mass-produced thin-film photovoltaic cells, the delivered cost of hydrogen would be on the high side—perhaps about $22–30/GJ, which is 3–4 times the benchmark price of $1 per gallon of gasoline. If all of the 14 million vehicles projected to be cruising the Los Angeles Basin in the early to mid twenty-first century were zero emitters, about 12 times that amount of land area—about 250 square miles—would be needed. Ogden et al. report there are several excellent wind sites, including the Tehachapi Pass and the San Gorgonio Pass, at the eastern edge of the Los Angeles Basin. They say that in the next 10 or 20 years hydrogen from wind will likely be less costly than photovoltaic-electric hydrogen, though in the long term they will probably be equal.

In general, Ogden et al. conclude, the delivered cost of hydrogen as a transportation fuel from existing sources could be as low as $12 and as high as $30 per gigajoule under the specific circumstances of the Los Angeles area. But once again it must be kept in mind that these higher per-unit energy costs are likely to be almost compensated for by the fuel cell's higher efficiency and lower maintenance costs. And much cleaner air for smog-saturated Angelenos would be the desired end result (and, let's not forget, the principal reason for moving to these technologies).

As a very rough rule of thumb, Ogden says that the chances for electrolytic hydrogen are probably best when one or more of these four things are true: The cost of electricity is very low (1–2 cents/kWh). The price of natural gas is relatively high (greater than $4 per million Btu. The demand for hydrogen is very small ("on the order of a few cars"). It is possible to make hydrogen on a small scale on site, thus avoiding the large costs of transmission and distribution.

There is a environmental downside to electrolytic hydrogen, though—if it is produced with commercial electric power. Sandy Thomas, Vice President for Energy and Environment at Directed Technologies, Inc., estimates that greenhouse gases in the United States would more than double if run-of-the-mill electricity were used to generate hydrogen, even if that hydrogen were to be consumed in efficient fuel cell vehicles. The reason is that more than half of the electricity in the United States is produced by burning coal. Using more coal-derived electricity to make fuel for cars and trucks would make a bad situation worse. "Hydrogen is not always a net benefit, and in this case can exacerbate climate change," according to Thomas. Even in California, which has proportionately more non-polluting nuclear and hydroelectric power sources than the rest of the country, hydrogen consumed in a fuel cell vehicle "would increase greenhouse gases by over 80 percent compared to conventional gasoline cars," Thomas said in a personal communication.

Primary Renewable Sources

Hydropower
Energy extracted from falling water is still the leading renewable energy source in the United States, according to an overview paper by Linda Church Ciocci, executive director of the Hydro Research Foundation.[9] It accounts for about 10 percent (about 90,000 megawatts) of the electric power generated in the United States, and almost 98 percent of the renewable energy.

Ironically, hydropower is growing more slowly than other forms of electricity production in the United States, according to the Hydro Research Foundation. Its share will fall to less than 7 percent by 2015. But in the world as a whole, interest in hydropower is rising. It accounts already for 19 percent of the world's net electricity generation and 24 percent of the world's capacity. Some 131,000 megawatts of hydroelectric capacity are planned for Central and South America, and another 127,000 MW for Asia. The biggest hydroelectric plant today (12,600 MW) is at Itaipu, on the Parana River in Brazil.

In terms of hydrogen production, the most esoterically ambitious hydroelectric project—one that is intended to be the first link to a future hydrogen economy—is proceeding in Iceland, where a parliamentarian named Hjalmar Arnason assembled a task force in the 1990s that essentially proposed converting the country's cars, its trucks, and its fishing fleet to fuel cells and hydrogen. (Iceland's exploitable hydroelectric reserve is estimated at 64,000 GWh/year, less than 10 percent of which is utilized.)[10] In early 1999 the Icelandic Hydrogen and Fuel Cell Company was formed, with three potent multinationals—DaimlerChrysler, Royal Dutch/Shell Group, and Norsk Hydro—as minority partners.

A coldly exotic variant of this primary energy source was suggested by Swiss researchers in the mid 1970s for Greenland. Ulrich La Roche, a scientist then working for the Swiss company Brown Boveri, proposed tapping Greenland's "glacier power," which by his estimate amounted to somewhere between 100 gigawatts and a terawatt[11]—far more than what the population in the ice-free coastal range would need for itself, and therefore exportable. La Roche presented his ideas at the 1976 Miami Beach hydrogen conference.

Wind

Next to hydropower, wind energy is generally believed to be the most cost-effective, least capital-intensive form of alternative primary energy.

In 1995 the installed wind power capacity in the United States was about 1750 megawatts. By mid 2000 this had increased to 2500 MW, with at least an additional 1000 MW expected to be developed by the end of 2001, according to Randall Swisher, Executive Director of the American Wind Energy Association.[12] Worldwide, more than 15 GW of wind capacity had been installed by mid 2000, and the total world capacity was doubling about every 3 years. Overall, Europe was the world leader, with almost 10 GW installed, and was expected to hold that position with the planned development of new offshore wind farms taking advantage of steady ocean winds. Remarkably, India was among the world's top five nations, with more than 1 GW installed. "Due to dramatic cost and performance improvements, wind energy is a reliable alternative that is more competitive than conventional power technologies such as coal, oil and nuclear energy," Swisher wrote in 1997.[13]

Swisher sees a combination of wind and natural gas as transitional: "Today, only natural gas is marginally cheaper than wind. Together, they will fulfill increasingly more of society's future needs, while minimizing the impacts on the environment." California's Tehachapi and Altamont Passes, blessed with strong daily wind cycles caused by temperature differences between land and water, are prime areas for wind power, says Swisher. Also potentially suitable, according to Swisher, are "a broad swath from northern Texas to North Dakota and Minnesota," the high plains of Colorado,

Wyoming, and Montana, and mountain ridges in the Appalachian and western mountain ranges.

A 1997 brochure put out by the American Solar Energy Society[14] says that wind power is cheap. In the United States, according to contributor Paul Gipe, the cost of a kilowatt-hour of wind-generated energy has dropped dramatically from more than 40 cents in the early 1980s to as low as 5 cents; according to Swisher, some utility contracts in Texas provide wind-generated electricity at levelized costs of 3 cents per kilowatt hour. That's good, but the costs of a kilowatt-hour of electricity will have to drop to below 3 cents, and perhaps as low as 1.5 cent, before electrically generated hydrogen can compete with gasoline.

Direct Solar Generation of Electricity

Probably the most successful demonstration of harvesting the sun's power to supply electricity to homes and industry involves nine solar stations in the Mojave Desert, between Los Angeles and Las Vegas, at Daggett, Kramer Junction, and Harper Lake, California. These plants, so far producing a total of 354 megawatts, use parabolic trough mirrors. The first went on line in 1984; construction of the last one ended in 1990. Designed and built by the American-Israeli firm Luz International Ltd., these plants have been converting solar power to electricity at peak efficiencies of about 21 percent, with an annual average efficiency of 12 percent and current costs of 8–10 cents per kilowatt-hour[15]—not cheap enough for hydrogen production.

Photovoltaics

Spurred by the oil shocks of the 1970s, the US government has been supporting the development of photovoltaics ever since. As a result, the prices of photovoltaic modules have dropped from about $100 per watt in 1975 to around $4 per watt.[16] During that time, the worldwide market for the technology has grown from just about zero to more than 90 megawatts—still tiny in comparison with the markets for wind and hydroelectric power.

The efficiency of the traditional crystalline cells (which are sawed laboriously out of pure silicone ingots) has increased from 4 percent to almost 25 percent for laboratory specimen; commercial production modules are about 14 percent efficient. The newer thin-film photovoltaic materials have lower efficiencies so far—17.1 percent for laboratory samples and 10 percent for commercial production modules. However, the thin-film materials can be produced as long ribbons, with much more automation than can be used in the production of crystalline cells, and that is expected to cut their manufacturing costs drastically. According to estimates by an industry association, the Utility PhotoVoltaic Group, the demand for photovoltaic systems in the United States may rise to 9000 megawatts once the cost drops below $3 per watt.[17]

Donald Osborn, supervisor of the solar program of the Sacramento Municipal Utility District in California, says that analysts expect the price of a single-crystal module to be as low as $2 per watt by 2010. Osborne says that advanced thin-film cells made of exotic materials such as gallium arsenide have shown laboratory efficiencies as high as 28 percent with concentrated sunlight.[18] Some thin-film photovoltaic modules are expected to cost as little as $1.25 per watt by 2010.

Biomass

Biomass (which includes, among other things, residues from wood harvesting such as bark and from pulp and paper manufacture, agricultural wastes, and gases generated in landfills from municipal solid wastes) can be burned with coal in conventional coal-fired power plants, burned separately in steam plants, or gasified to run gas turbines, fuel cells, or internal-combustion engines.[19]

Biomass electric plants typically use small steam turbines (10–30 MW; some go as high as 50 MW), because usually the amount of biomass at a given site is relatively small. Since these plants are small, they cannot economically incorporate sophisticated heat-recovery components, and their efficiency is therefore relatively low: 17–23 percent, with some larger ones achieving 28 percent. (A large modern coal plant gets up to about 35 percent.) Still, direct-combustion biomass plants produce electricity at fairly low cost—about $0.073/kWh.[20]

Higher efficiencies can be achieved by first gasifying biomass and then running it through a high-efficiency gas turbine system. Systems that do this (now under development, primarily for coal) are expected to be quite cost-effective, making electricity at somewhat lower costs—about 7.1 cents per kWh. By 2020, electricity from gasified biomass is expected to cost only about 4.3 cents per kWh.

One promising method may involve using marginal or underused lands to raise fast-growing energy crops. Three demonstration "crops-to-power" projects sponsored by the US Energy and Agriculture Departments were begun in the mid 1990s.[21] One of them, in western Minnesota, burns Alfalfa stems (about 50 percent of the crop); the leaves are used for animal feed. A second involves growing so-called switchgrass on 40,000 acres of marginal cropland in Iowa to provide 35 megawatts of co-fired power in a 735-MW power plant. (Switchgrass is a deep-rooted indigenous prairie grass that helps to build up and stabilize the soil, efficiently reseeds itself, and it doesn't need pesticides.) A third, led by the Niagara Mohawk Power Corporation, involves the cultivation of willow trees as a fuel supplement to coal-fired power production. A consortium planned to grow willows on plantations located in a 50-mile radius from selected coal-burning power plants in upstate New York. Willows can reach heights of 28 feet in 3 years, after which they can be harvested with no need for reseeding; the consortium expects seven harvests in 21 years.

Biomass does produce emissions, including nitrogen oxides, particulates, volatile organic compounds, and various toxics. It is argued, however, that these pollutants are well understood and treatable, and that emissions should meet air-quality standards.[22]

But the major attraction of biomass energy, from the perspective of reducing greenhouse gases, is that it is carbon-neutral: the CO_2 generated when biomass is burned will be captured again with new plant growth—an elegant match with the natural carbon cycle. Critics point out, though, that net carbon emissions do result if the plants are not regrown. Also, farm equipment, trucks, and other machinery needed to harvest, transport, and process biomass all use fuels. If those fuels are gasoline or diesel, carbon emissions result. If they are biomass-derived, they cut overall biomass energy production.

Advanced Solar Concepts

Coming up behind the almost-ready-for-industrial-prime-time technologies are new technologies that tap into solar power through novel (and, it is hoped, more efficient) ways to make electricity or hydrogen. One of them is direct high-temperature solar water splitting, literally the hottest and technically one of the most difficult methods of solar hydrogen production. The basic idea is to use highly concentrated sunlight, focused by a large number of ground-level mirrors onto a point atop a solar tower, to generate temperatures of thousands of degrees, at which steam dissociates into hydrogen and oxygen.

Attempts in France, Canada, and Japan in the late 1970s and the early 1980s to split water directly using focused sunlight typically converted only about 10 percent of the steam into its constituent gases. More recently, H Ion, a small company based in Richmond, California, recorded conversion of only 2–3 percent, with "momentary" readings of 10 percent, in an effort initially supported by the National Renewable Energy Laboratory.

A team at the Weizmann Institute of Science in Israel, headed by Abraham Kogan, tackled the problem again in the early 1990s, believing it could do much better and perhaps even develop an industrially viable process. In a 1996 interview, Kogan explained that previous efforts had attempted direct thermal water splitting at 2500°K (2227°C) and normal atmospheric pressure, and noted that only about 10 percent of the water had been dissociated into hydrogen and oxygen.[23] "We intend to operate at 2500°K and 0.05 atmosphere, a state at which 25 percent of the water is split," he said. The Weizmann team ran its first tests at the institute's 64-heliostat, 3-megawatt solar tower in March 1993. First traces of hydrogen were detected at 1650°C. At 1750°C, hydrogen began to flow at a rate of 30 milliliters per minute, Kogan reported at a 1995 conference. Eventually, Kogan expects high enough yields to make direct solar water splitting economically feasible. According to a Weizmann Institute video on the project, splitting of up to 30 percent of steam in the solar reactor at 2300°C, and perhaps even 55 percent at 2500°C, is achievable.

During the 1970s, in the aftermath of the early oil shocks, a number of imaginative alternative energy schemes popped up in various places. Some examples follow.

Giant Solar Chimneys

Large-scale production of electricity from solar power is the goal of the "solar chimney" idea proposed by two West German engineers, Jörg Schlaich and Rudolf Bergermann, as early as 1976. The basic idea is simple. It combines the greenhouse effect (air and soil are heated underneath the translucent collector roof by solar radiation) with the chimney effect (a strong upward air draft created by the temperature differential drives wind turbines and power generators). Such plants, says Schlaich, could produce electricity at competitive costs.

Schlaich and Bergermann envisioned a huge hybrid industrial smoke-stack and cooling tower, up to 3300 feet high and 330 feet or even larger in diameter. The chimney would be built of lightweight concrete or would be assembled from cable mesh and aluminum cladding hung from a central concrete pillar. At ground level, the chimney would be surrounded by a translucent roof of glass or plastic, which would be 6–18 feet above the ground and up to about 7 miles in diameter. As an added benefit, the roof's peripheral areas, where the air mass drawn to the center would be only a gentle breeze, could be used to grow crops.

Based in part on German utility data, Schlaich calculated that a 400-megawatt solar chimney plant would at first produce electricity at a cost of DM 0.14 per kilowatt (8 cents at mid-1998 exchange rates), versus DM 0.12 (7 cents) for a coal-fired plant and DM 0.1035 (6 cents) for a combined-cycle plant. These costs, he predicted, would drop by about two-thirds over the 60-years-plus life of the solar chimney.[24]

The first such solar chimney plant, mentioned at the beginning of this chapter, was completed in 1983 near Manzanares, about 100 miles south of Madrid. It ran successfully until 1990. In 1996, planning started for a considerably larger plant (200 MW) in the Indian state of Rajasthan. This plant, a collaborative effort of the Sri Lanka-based Energen International Consortium, Schlaich's office, and a team of international specialists in the construction of very tall structures, was projected to cost $800–900 million.[25] Energen's web site says this will be the first phase of a 1000-megawatt solar project. The 3300-foot tower will have a diameter of 557 feet, and the roofed "greenhouse" collector area on the ground will have a diameter of almost 7 miles.[26]

A possible drawback of solar chimneys is that their efficiency in converting solar energy to electricity energy is fairly poor. For the 200-megawatt plant planned for India, Schlaich's office projects an efficiency of only about 3 percent—a lot less than the double-digit efficiencies claimed for the trough solar plants in the Mojave Desert.

Orbiting Solar Mirrors
An idea that received quite a bit of media attention in the 1970s was to collect solar energy by means of orbiting solar power stations, which would then transmit huge quantities of electric power to Earth in microwave

beams. In an alternative scheme, an orbiting system of mirrors would reflect and focus sunlight onto Earth-based solar receiver power plants, which would then convert the solar energy into electricity.

Boeing Aerospace and Rockwell International were said to be exploring the first idea. NASA and the Department of Energy spent some money on a multi-year feasibility study. A 1978 article in the magazine *Engineering News-Record* quoted Peter Glaser, then president for engineering at Arthur D. Little, Inc., as saying that such systems could supply as much as 25 percent of the planet's energy requirements by 2025.

The second approach—putting a number of mirrors in orbit to beam sunlight onto an "energy oasis" about 25 miles in diameter—was outlined in 1977 by Krafft Ehricke, a former NASA manager.[27] Ehricke said that "an industrial sun for Europe" would create a climate in the "energy oasis" roughly equivalent to that of the Arab or the Australian desert, and that the total yield from the "oasis" would be between 35,000 and 50,000 megawatts. Obviously, the idea never flew.

Converting Thermal Energy from Oceans

Exploiting the temperature differences between warm near-surface layers of the oceans and deeper, colder layers is a technology whose time never really came. Despite intermittent high hopes over the decades, ocean thermal energy conversion (OTEC) seems to have been relegated to oblivion. A French physicist, Jacques d'Arsonval, first suggested the basics more than 100 years ago. In the early 1930s, a French engineer, Georges Claude, built an ocean-thermal plant on the shore of Matanzas Bay, about 50 miles west of Havana, that produced about 22 kilowatts of electricity. Claude also built a plant aboard a cargo vessel moored off the coast of Brazil. Both plants were eventually destroyed by weather and waves, and neither appears to have produced more electricity than was needed to run it. Cheap fossil fuels contributed to their demise.

About 25 years later, French researchers designed a 3-megawatt OTEC plant for Abidjan (on Africa's west coast), but the plant was never completed. In 1979, the Natural Energy Laboratory of Hawaii (at Keahole Point, on the Kona coast of the island of Hawaii) built a 50-kilowatt plant on a Navy barge a short distance offshore. The plant produced 52 kilowatts of electric power, but once again most of the power was needed to run the system—

the net output was only 15 kW, according to the National Renewable Energy Laboratory's OTEC home page. Another OTEC plant was built in 1981 by Japan on the island of Nauru. That 100-kilowatt plant generated about 32 kilowatts net power. And as recently as 1993, an "open-cycle"[28] plant at Keahole Point produced 50 kW of electricity in an experiment.

The international hydrogen community was introduced to OTEC power at the 1974 Miami Beach THEME conference with a paper that envisioned the conversion of solar energy via both photovoltaic collectors and OTEC plants mounted on large floating platforms. Nothing ever came of that vision. The US Department of Energy funded an OTEC program in the 1970s, but there were doubts about its feasibility almost from the start.

Except for the facilities in Hawaii (now run by the Pacific International Center for High Technology Research in Honolulu), OTEC has almost faded away, reported Luis Vega, a PICTR researcher, in the autumn of 1997. The DoE has stopped funding it. "Presently there is no OTEC activity in the world," according to Vega.[29]

Geothermal Energy
Geothermal energy may be the real sleeper—and a giant one at that— among all the options for sustainable energy sources. Geothermal electricity has been produced continuously since 1904 in the venerable Larderello geothermal fields in Italy's Tuscany region, some 40 miles southwest of Florence, according to a 1998 brochure ("Geothermal Energy") issued by the University of Utah's Energy and Geoscience Institute. Another 1998 brochure, this one produced by the DoE's Office of Geothermal Technologies, says that geothermal electricity has been generated without interruption in New Zealand since 1958 and at The Geysers field in California since 1960. "In fact, no geothermal field has been abandoned because of resource decline," according to the latter document.[30]

Today, geothermal energy produces about 2700 megawatts of electrical power in the United States. The world's total output is about 7000–8000 MW, according to the DoE. Geothermal power plants operate at high capacity factors (70–100 percent), and their availability factors typically exceed 95 percent.

The total geothermal resources of the United States are estimated at 70,000,000 quads, "a truly enormous amount of energy," according to the

DoE brochure—the equivalent of a 750,000 year supply of energy at current consumption rates. The University of Utah brochure says that 10,000–15,000 megawatts of electric generation is a "reasonable" estimate of what could be developed in the United States, and that 50,000–100,000 MW is deemed possible worldwide from "hydrothermal" resources alone. Overall, geothermal energy ranks as the third largest renewable energy resource, behind hydroelectric power and biomass and exceeding solar and wind energy, according to the Utah estimates. With research now underway to find and exploit hidden energy resources such as hot springs (and, ultimately, dry hot rocks in volcanic-magmatic layers at depths of more than 10 miles), these energy resources will eventually be truly huge. Indeed, they are believed to be much larger than the combined resource bases of coal, oil, gas, and uranium.

A belt of geothermal "provinces" rings the Pacific Ocean, curling southward from the Indian subcontinent through Southeast Asia to Indonesia and then sweeping northward again along China's and Japan's shores, jumping to North America via the Aleutians and then stretching southward again along the western coasts of Canada, the United States, and South America. A second, smaller such "province" stretches from the Straits of Gibraltar across the Mediterranean and into Northern Italy, Yugoslavia, and Russia, with a southern extension hugging the Red Sea and stretching deep into eastern Africa. And there are smaller geothermal hot spots in Iceland, in the Canary Islands, and in the Hawaiian Islands.

Geothermal energy is clean. According to the aforementioned DoE brochure:

- No fuel is burned, there are no nitrogen oxide emissions, and very low sulfur dioxide emissions
- No air emissions with binary[31] geothermal plants
- Many geothermal plants generate no appreciable waste
- Plants have small land use over the project's lifetime, and low impact in scenic regions.

"Geothermal energy," the University of Utah document concludes, "is a vital part of a sustainable future."

6

Hydrogen for Cars and Buses: Steaming Tailpipes

"It doesn't happen every day that a politician waxes euphoric at the sight of a gas station pump," the respected German daily *Süddeutsche Zeitung* observed on January 13, 1999. But that's just what Mayor Ortwin Runde of Hamburg did at the opening of Europe's first commercial hydrogen refueling station: "The streets will be quiet. Only the sound of tires and rushing wind will accompany passing vehicles instead of the roar from exhaust pipes. The city will be clean, since emissions will be practically zero. Pedestrians strolling on the sidewalks won't be turning up their noses, guests won't be fleeing from the street's stench into the cafes because now they can enjoy their sundowners in the open air." Runde was the first to take the wheel of the first of six delivery vans converted to gaseous hydrogen for a ceremonial spin. The van, a 3-ton, 2.3-liter, 4-cylinder internal-combustion-engine Mercedes-Benz, belongs to a delivery fleet operated by the mail-order company Otto-Versand, one of the world's biggest. (In the United States, Spiegel and Eddie Bauer belong to the Otto Group.)

Otto-Versand has been promoting environmental responsibility and sustainability in its products for years. It began experimenting with natural gas-fueled vans in 1995, and in 1998 it installed a 50-kilowatt photovoltaic system—the biggest in Northern Germany—at one of its facilities. In 1997 it teamed up with a dozen local companies on the hydrogen gas station project initiated by the Hamburg Hydrogen Society, a nonprofit group founded in the 1980s.

The participants in the project include two banks, Hamburg's transit authority, the city's electric and gas utilities, a gas distributor, a specialized moving company, and, notably, Deutsche Shell AG, the German division of the Royal Dutch/Shell Group. The van's conversion to hydrogen operation was designed, built, and tested by a veteran American hydrogen specialist, Frank Lynch of Hydrogen Components in Littleton, Colorado.

In addition to wide coverage in regional and national papers, all of Germany's TV stations and CNN showed footage of the van being refueled and driven in and out of the "Ludwig Boelkow Fueling Station," named after the aerospace industrialist who was the patriarch among Germany's hydrogen supporters. Even the *New York Times* carried a brief story from the Reuters wire service. "Hydrogen will be the most important energy source of the 21st century," Fritz Vahrenholt, a top Deutsche Shell executive, was quoted as saying. "Long term, it will replace oil and gas."

Initially, the Otto-Versand vans were to be refueled by commercially produced hydrogen at costs somewhat higher than those of conventional fuel—about 13 cents per kilometer, 4 cents more than diesel fuel. But the goal of the multi-year WEIT (Wasserstoff-Energie Island-Transfer, meaning Hydrogen-Energy Iceland Transfer) effort is to import renewably produced hydrogen by ship from Iceland, where equipment in an old fertilizer plant is used to make hydrogen from renewable sources such as geothermal and hydroelectric power.

Mindful of the wide publicity and the high hopes that fuel cell technology has garnered in recent years, the organizers of the Otto-Versand project emphasized that the goal was not to develop and build the most sophisticated, best possible hydrogen power system. Rather, they said, the goals were to demonstrate that hydrogen vehicles can operate reliably in a normal urban business environment and to create public acceptance. The vans are driven by company pool drivers with no special training or licensing, and the station is designed for do-it-yourself refueling.

To Michael Otto, CEO of the company that bears his name, the January event was profoundly significant. Getting the first hydrogen-fueled van on the road was "not going to be a one-time flash in the pan," he said in his remarks during the ceremony. "The Hamburg hydrogen vehicle fleet is emblematic of the course toward a future-oriented utilization of hydrogen energy in road transportation." He continued: "As business executives, it is our task to think systematically about the future, to proactively effect change and to shape the future. I am deeply convinced that it is no longer enough to look only at the profit-and-loss ledger. Only those entrepreneurs who look at the total balance sheet and who are conscious of their responsibilities toward society and the environment will be successful in the long run. The energy supply of our planet is a task of existential significance for mankind. Obviously, such investments are costly in the short term. But in the long term it will become clear that ecology and economy are not necessarily opposites. On the contrary, investments in environmental technologies of the future will assure competitiveness and new jobs."

Two months later, in Washington, DaimlerChrysler made headlines around the globe with the unveiling of its latest fuel cell car, NECAR (New Electric Car) 4. DaimlerChrysler co-chairman Bob Eaton drove the tiny four-seater onto center stage at a jammed press conference in the Ronald Reagan International Trade Center. The LH_2-powered NECAR 4 (which is based on the A-Class four-passenger subcompact, sold only in Europe) has a top speed of 90 miles per hour and has a range of almost 280 miles. The twin fuel cell stacks put out 70 kilowatts. NECAR 2, a liquid-hydrogen-powered mini-van larger than NECAR 4, had a top speed of only 68 mph. (NECAR 3, also derived from the A-Class, was fueled by methanol, DaimlerChrysler's avowed main fuel preference for a commercial fuel cell car.)

Lest anyone doubt that DaimlerChrysler was into fuel cells for keeps, co-chairman Jürgen Schrempp said at the Washington festivities: "Today, we declare the race to demonstrate the technical viability of fuel cell vehicles is over. Now, we begin the race to make them affordable." The company plans to have fuel cell vehicles in limited production by 2004, Schrempp said. Toward that end, it intends to spend more than $1.4 billion on fuel cell technology, he added—about the same amount of money invested in an entire line of profit-making vehicles (such as the Dodge Intrepid and its derivatives) in the United States. "NECAR 4 is a major breakthrough in fuel cell technology because we've developed a very powerful fuel cell system small enough to be packaged in a compact car," added Ferdinand Panik, head of DaimlerChrysler's Fuel Cell Project. "Five years ago, you needed a large van (NECAR 1) to contain all the fuel cell hardware. Now, that's simply no longer an issue. The significance of this technological advancement is comparable to the impact the microchip had on computer technology when it replaced the transistor."

Otto and Schrempp were emblematic of the change of attitude among some international business leaders who came out in favor of advanced hydrogen-based energy technologies as the twentieth century drew to a close. Key executives from just about all major auto manufacturers have said about the same. "Looking at the total system, including its packaging and incorporating the fuel cell and hydrogen absorbing alloy, we've achieved what I believe to be the world's highest standard," said Yoshio Kimura, Toyota's general manager for the experimental fuel cell RAV4 project, at the car's unveiling at the Osaka International Electric Vehicle Symposium in October 1996. And John Smith Jr., chairman of the board of directors of General Motors, predicted at the 1997 Detroit Auto Show that "environmental pressures will force changes in the automotive industry and lead to a fundamental shift away from gasoline-powered vehicles to electric, hybrid, and fuel-cell-powered vehicles," and that "no car company will be able to thrive in the future with 100 percent dependency on internal-combustion engines."[1]

This was a profound change in attitude. Through the 1970s, and probably until the late 1980s, any talk of advanced propulsion technologies, clean energy, or concern with environmental issues was considered heresy in the blinkered automotive world of Detroit. William Clay Ford Jr. got a taste of that when he tried to raise environmental issues after joining the company founded by his great-grandfather, Henry Ford, as a vehicle planning analyst in 1979. "Coming to an old-line auto company, people looked at me like I was a Bolshevik for bringing it up, for even asking the questions," Ford told a *New York Times* reporter in early 1999.[2] Later, when he was Ford's director of business strategy in the early 1980s, Ford sent out a memorandum calling a meeting on environmental policy, only to be warned immediately by company lawyers that even the memo, not to mention the meeting, could be used against Ford in an environmental lawsuit, according to the *Times* story. On several occasions William Ford was at the point of quitting the company and starting an environmentalist group, only to be told by his wife and friends that he could have more of an impact if he stayed in the firm. "And they were right, ultimately, but they were more patient than I was," the story quoted Ford (now the company's chairman) as saying.

There is no longer any real question that advanced transportation technologies employing hydrogen (directly or indirectly) as an energy medium will be essential in the twenty-first century. That is a far cry from the hesitancy, uncertainty, disbelief, ironic skepticism, and raised eyebrows that accompanied discussions about hydrogen as late as the mid 1990s.

The case for alternative transportation fuels in general and hydrogen in particular has been argued for years in all sorts of forums. The intent here is not to plow the same ground again in great detail. Still, some numbers are pertinent.

In 1997, the last year for which estimates were available at the time of this writing, transportation accounted for about 54 percent of oil consumption in the United States (that is, about 18.6 million barrels per day).[3] More than half has to be imported, much of it from the politically unstable Middle East. Transportation is arguably the sector of society that contributes most to air pollution. It accounts for 76.6 percent of US carbon monoxide emissions from all sources, 49.2 percent of nitrogen oxide discharges, 40 percent of volatile organic compounds, and 23 percent of particulate matter, according to the Environmental Protection Agency.[4] Carbon dioxide accounts for about 84 percent of US greenhouse-gas emissions from all human activities, according to estimates by the DoE's Energy Information Administration for 1997 (the last year for which data are available). Ominously, emissions per dollar of GDP and per capita, which had declined in the early 1980s, rose in the 1990s, although at a low rate. Transportation accounted for about one-third of CO_2 emissions, the Energy Information Administration reported. Total CO_2 emissions in 1979 were 473 million metric tons.[5]

Hydrogen is widely regarded as the ideal or nearly ideal fuel to solve those problems. By definition, it doesn't pollute. Burned or oxidized with atmospheric oxygen, it produces water. (It also produces some nitrogen oxides if combustion occurs with a flame, as in an internal-combustion engine, but no nitrogen oxides are produced in an electrochemically reactive fuel cell.)

The difficulty lies in how to carry hydrogen onboard a vehicle. Hydrogen is difficult and unwieldy to store on any vehicle, and especially on a space-constrained passenger car. Carmakers and researchers have wrestled with this for decades, and there is still no clear solution in sight. Will hydrogen be carried in a metal hydride (as in Toyota's experimental RAV4, and in the 1980s in a Mercedes fleet test)? As a compressed gas in tanks (as done in the experimental Daimler-Benz NECAR 2, in fuel cell concept cars under development at Ford, and in several fuel cell buses recently launched by DaimlerChrysler, MAN Nutzfahrzeuge, and Neoplan)? As a cryogenic liquid (BMW's preference)? Or will it be extracted from a carbonaceous fuel such as gasoline (an approach investigated by most carmakers as a backup, and also by the Department of Energy), or from methanol (the approach that DaimlerChrysler thinks is most likely to be commercially successful)?[6]

In gasoline- or diesel-powered cars, storing conventional liquid fuel is easy. Conventional front-engine designs usually have the gas tank in the rear, beneath the trunk. With hydrogen, the situation is more complicated. In its ambient state, hydrogen is a gas. For efficient storage, it must be either compressed like natural gas, cooled into a cryogenic state, bound within the structure of a hydride, or put into some other form (e.g., a slurry). All of the above are difficult and/or have engineering drawbacks and/or economic drawbacks. As the twenty-first century dawned, there was no universal agreement on how best to store hydrogen onboard a vehicle. That is why some researchers and some carmakers are looking at liquid hydrocarbon fuels such as methanol and various types of hydrocarbon fuels (including gasoline and diesel) as "hydrogen carriers." "We badly need a breakthrough," the Department of Energy's hydrogen program manager, Neil Rossmeissl, told a 1996 meeting in Miami at which the DoE's hydrogen program was reviewed. The situation remains basically unchanged.

No big carmaker has made a firm decision as to what storage method it will use in a commercial vehicle. Makers of hydrogen equipment (e.g., the Electrolyser Corporation and its recently formed transportation subsidiary, Stuart Energy) and researchers (e.g., at Princeton University's Center for Energy and Environmental Studies) argue that building a pure hydrogen infrastructure "up front" would not be much more costly overall than installing a hydrogen-extracting micro-plant under the hood of each of millions of future fuel cell vehicles. In fact, Joan Ogden thinks that until a large demand for hydrogen is built up, production of hydrogen at fueling stations itself or truck delivery will offer the lowest delivered fuel costs.[7] In the long term, hydrogen might be made at the city from the best local resource and distributed via local pipelines to refueling stations. Onboard storage of compressed hydrogen gas, at pressures of up to 5000 pounds per square inch, has recently been getting increasing attention at the DoE and among carmakers, and a few manufacturers (including IMPCO Technologies and Dynetek) are beginning to offer such systems.

Purists decry the hydrogen-from-carbon-fuels strategy as a betrayal and a sellout to Big Oil, but proponents say that—at least as an interim solution until a better hydrogen-storage technology comes along, and in the absence of an extensive hydrogen refueling infrastructure—this may be the way to go. This would resolve the decades-old chicken-and-egg dilemma (what should come first, a power plant tailor-made for hydrogen or a hydrogen supply system?), permitting fuel cells to be phased in with minimal infrastructure dislocation and societal cost. By the time an efficient onboard hydrogen-storage method becomes practical, the argument goes, fuel cell power plants will be widespread, and their adaptation to pure hydrogen may be relatively painless.

Back in the 1970s, when people first thought about using hydrogen as fuel for cars and trucks, many assumed that heavy steel-walled pressure bottles, similar to those used in welding, would be used to store the fuel onboard. Such bottles are simple, but their weight would be prohibitive. Larry Williams of Martin Marietta Aerospace once calculated that a conventionally constructed steel pressure tank capable of holding roughly the same amount of energy as the fuel tank of a standard-size car would have to weigh about 3400 pounds, would require a pressure of 800 atmospheres, and would have to have steel walls almost 3 inches thick.

Larry Williams, whose advocacy of cryogenic storage has not wavered in 20 years, listed the following main advantages of this storage method in a 1973 paper:

lowest cost per unit energy
lowest weight per unit of energy
simple supply logistics
normal refuel time required
no unsurmountable safety problems.

Williams listed these disadvantages:

loss of fuel when vehicle is not in operation (the supercold liquid evaporates)
large tank size
cryogenic liquid engineering problems.

In a personal communication, another expert, Sandy Thomas, adds this disadvantage:

Liquid hydrogen derived from natural gas used in a very efficient fuel cell would not reduce greenhouse gases significantly. Again, coal-produced electricity would be the culprit: the electricity requirement to liquefy hydrogen, produced by burning coal, approximately offsets the improved fuel economy of the fuel-cell vehicle.

Some of the problems listed by Williams—large tanks, engineering problems, boiloff—have been minimized by the two main developers of liquid-hydrogen transportation technology in the last two decades, BMW and the Musashi Institute of Technology. (N.B.: Hereafter, "liquid hydrogen" will often be abbreviated "LH_2.")

A paper presented by BMW at the 1996 World Hydrogen Energy Conference in Stuttgart illustrated progress with photographs of the tanks in the trunks of BMW's four generations of LH_2-fueled cars. In the first picture, which showed the original 1979 model, the tank, with protruding pipes and fittings, occupied most of the trunk space. The 1984 version was a smoother, ovoid-shaped structure that still occupied much of the space. The 1990 version was a smaller cylindrical structure that left much more room for luggage. In the 1995 version, the tank was so intelligently packaged and incorporated in the trunk above the rear axle as to be almost invisible, leaving lots of useful space.

Another paper presented at the Stuttgart conference—this one from Solar-Wasserstoff-Bayern, a now-defunct experimental solar hydrogen facility near Munich—described a 3000-liter LH_2 filling station it had built to study and optimize terrestrial refueling techniques for the rocket fuel. It described in detail how the sophisticated coaxial fuel line couplings and cryogenic onboard tanks, developed competitively by two German industrial gas suppliers, Linde and Messer-Griesheim, had made refueling easy. Those experiments culminated in the mid-1999 unveiling of a showcase fueling station dispensing both liquid and gaseous hydrogen at Munich's airport. (As part of the development process before the airport station was unveiled, some 500 filling tests, equal to more than 90,000 miles of road tests, were conducted, using BMW sedans.)

A third paper, presented by Messer-Griesheim, announced that boiloff losses of LH_2 had been reduced from about 9 percent per day 20 years earlier to about 1 percent. "Actually, we experience no fuel loss during normal operation," Messer-Griesheim reported, "because the pressure holding time could be increased to several days." And the energy density for LH_2 tanks has been increased over several generations of tank designs from 10 to 22 megajoules per kilogram with the use of better, lighter insulation materials and with other improvements.

For a long time, hydrides were considered to be the perhaps best solution to the problem of how to store hydrogen in an automobile. Safety was the main reason. Ever mindful of the "Hindenburg syndrome," researchers generally consider hydrides safer all around, because they cannot spill or vent hydrogen and they do not burn in a crash. In recent years, however, the weight penalties associated with hydrides have dampened the enthusiasm for them considerably. Daimler-Benz, an early advocate, has dropped hydrides in favor of methanol as a hydrogen carrier for fuel cell cars, and in favor of compressed hydrogen gas for buses. Still, Mazda experimented with hydrides in the 1990s. In 1996, Toyota was apparently the only car-maker to employ hydride storage (in the experimental fuel-cell-powered RAV4 it unveiled that year). Two niche applications in which hydride storage is still considered viable are fork-lift trucks and mining vehicles. And in early 2000, Energy Conversion Devices Inc. of Troy, Michigan, was touting its magnesium-based, "atomically engineered" hydrides as capable of storing hydrogen efficiently and economically.

Hydrides—usually metal alloys[8]—were originally developed for nuclear power plants to slow down fast neutrons as a means of controlling nuclear reactions and thus controlling the output of a nuclear power plant. Titanium-iron hydride, the material used in early automotive applications, looks and feels like any ordinary metal—tiny silvery granules without any hint of the unusual. However, the granules have the remarkable character-istic of absorbing hydrogen in an almost sponge-like fashion, with differ-ent hydrides soaking up and releasing the gas at different pressures and temperatures.

Heat is given off when the alloy or other storage material absorbs hydro-gen (heat of formation); the same amount of heat must be added to the hydride to release the hydrogen again (heat of decomposition). Hydrides bind the hydrogen atom atomically; the hydrogen atom is integrated into the crystalline structure of the hydride, with the hydrogen atom's electron transferred to the hydride.

Integrated into the alloy storage material, hydrogen does not take up any additional volume, as a gas or a liquid would. For this reason, any hydride can carry much more hydrogen energy in terms of volume than liquid hydrogen, but there is a huge weight penalty. For example, a 100-liter tank of titanium-iron hydride carries 1.2–1.5 times as much energy as a 100-liter tank of liquid hydrogen, but it weighs about 25 times as much.

Much of the early groundbreaking research on storing hydrogen in hydride alloys was done in the 1960s at Brookhaven National Laboratory in the United States and at Philips Laboratories in Holland. The first papers were presented in 1966. In 1969, a team of six scientists presented a study titled "Metal Hydrides as a Source of Fuel for Vehicular Propulsion" at the International Automotive Engineering Congress in Detroit.[9] Since then, researchers have been trying to find better, lighter hydrides that can store more hydrogen fuel, but the quest has been elusive. Although the so-called AB5 hydrides (developed by Philips, and based on alloys of nickel and lanthanide elements) evolved into today's ubiquitous nickel metal hydride batteries, there have been no major improvements beyond the 1970s' titanium-iron (TiFe) materials for vehicular storage of gaseous hydrogen, explained Gary Sandrock, an internationally recognized expert on hydride storage technology. AB2 alloys, typically containing titanium, zirconium, manganese, vanadium, and chromium, have somewhat better weight efficiency, as do some vanadium alloys; however, Sandrock says, "all ambient temperature metal hydrides are still considered too heavy for most vehicles. Those alloys that have acceptable weight efficiencies, typically based on magnesium, require undesirably high temperatures to release the hydrogen."[10]

In the late 1990s, Energy Conversion Devices began to successfully demonstrate metal hydride storage systems for hydrogen, as well as metal hydride batteries. At the February 2000 National Hydrogen Association meeting, ECD's Rosa Young described efforts to develop magnesium-based hydride hydrogen-storage systems for transportation with one of the highest reversible hydrogen-storage capacities—theoretically 7.6 weight percent—with low materials cost at relatively high temperatures of 300–400°C and pressures of less than 20 bars. So far ECD's systems suffer from poor kinetics, short cycle life, and high heat of formation, but ECD's Ovonic division is tackling these problems with what it calls an "atomic engineering approach."[11]

Thus, the benchmark for efficient onboard energy storage is still gasoline. The average tank, full of gas, weighs about 110 pounds. A tank full of methanol would weigh about the same, but would give only about half the range.

Interest in methanol as an alternative fuel revived in the late 1990s—for fuel cells. It wasn't the first time this type of alcohol attracted the attention of the clean-energy community. A number of strategists had argued decades earlier that hydrogen, normally a gas, might not be the preferred eco-fuel for automotive use after all and that perhaps some other liquid fuel might be preferable.

Methanol came up at the first international hydrogen conference, the 1974 THEME conference in Miami. A group of researchers from the Stanford Research Institute said that replacing gasoline as a fuel was very difficult because the "distribution network proves to be a dominant component of the total private vehicle transportation system." "Once established," they noted, "the infrastructure networks of a system become very resistant to change. . . . Institutional change is often less readily accomplished than technical change." Societies get locked into a given system. The paper concluded that "compared to several alternatives, especially the use of methanol in vehicles, a transition to hydrogen would appear to be needlessly disruptive."[12]

Methanol, a particular type of alcohol, was supported as the clean liquid fuel for cars, notably by Volkswagen in the early 1970s. Volkswagen began thinking about methanol in earnest in 1973. In 1975 it built a fleet of 40 methanol burners, which it deployed in various German cities. Ernst Fiala, Volkswagen's research chief at the time, explained that, after analysis of many factors, "it becomes evident that methanol is a very good fuel. . . . Based on our knowledge today, it is probable that in 50 years we'll be driving on methanol, more probably so than any other fuel."[13]

Methanol, a liquid fuel, is closest to hydrogen in terms of environmental cleanliness. It has only one carbon atom in its structure. It has been described as "two molecules of hydrogen gas made liquid by one molecule of carbon monoxide," and as thus sharing many of the virtues of pure hydrogen. Also called methyl alcohol, wood alcohol, or methylated spirits, methanol (CH_3OH) is a clear, odorless liquid that freezes at $-144°F$, boils at $148°F$, and mixes easily with water. In fact, it attracts water, which is one of the drawbacks when it comes to burning it in internal-combustion engines. But future direct-methanol fuel cells—fuel cells that will react methanol "straight," without needing a separate reformer to extract hydrogen—will operate on methanol-water mixtures.

Methanol can be made from many sources, including natural gas, petroleum, coal, oil shale, limestone, and even wood, farm, and municipal wastes. This is an important consideration: methanol, made from renewable plant matter, would be a greenhouse-gas-neutral fuel, since it would not add to the world's total CO_2.

One of the chief drawbacks of methanol, and one that has essentially precluded its wide use in internal-combustion engines, is its low energy content.

Per unit volume, it has only a little more than half the energy content of gasoline—64,700 Btu per gallon, versus 120,000 for gasoline. In practical terms, this means that a methanol-fueled car gets only 55–60 percent of the mileage of a conventional i.c.-engine car. Another problem is that methanol makes for hard starting in cold weather, because it lacks certain volatile components found in gasoline.

Another drawback that has a lot of people worried about its widespread introduction as a transportation fuel is that methanol is toxic, both as a vapor and as a liquid. A "material safety data sheet" published by Canada's Methanex Corporation, a global supplier of the chemical, says that "swallowing even small amounts of methanol may cause blindness or death" and that "effects of lower doses may be nausea, headache, abdominal pain, vomiting." A 1997 US government monograph titled *Methanol Toxicity*[14] warns on its cover page that "the shift to alternative motor fuels may significantly increase both acute and chronic methanol exposures in the general population." About 35,000 gasoline ingestions are reported annually in the United States, the report says, most of them from fuel siphoning— and "siphoning accidents could significantly increase the number of methanol ingestions if the use of methanol-containing automotive fuels is widespread."

Just about all of these problems and difficulties would disappear with the advent of a better onboard hydrogen-storage system. Beginning in 1997, it looked as if such a system was almost at hand. The international hydrogen community was set abuzz by reports that two scientists at Northeastern University, Nelly Rodriguez and Terry Baker, had developed a graphite nanofiber storage material that would store as much as 65 percent (by weight) of hydrogen. That was a truly sensational number; most conventional storage systems store hydrogen in low single-digit percentages. Rodriguez and Baker calculated that their system would soak up enough gaseous hydrogen at moderate pressures and room temperatures to give a fuel cell car a range of 5000 miles or so. A tank filled with graphite nanofibers would be about half the size of a conventional gasoline tank and would weigh about 190 pounds. The US Department of Energy and DaimlerChrysler supported Rodriguez and Baker for a while, but they stopped after concluding that nanofiber storage was not likely to be a commercial reality in time for the launch of commercial fuel cell cars planned for around 2004 or 2005. (Different manufacturers give different dates.) Apparently no other experts had been able to duplicate Rodriguez and Baker's astonishing claims. Skepticism ran rampant. Still, Rodriguez and Baker weren't giving up. Subsequently supported by Ford and perhaps others, they continued their investigations. At a Canadian hydrogen conference in early 1999, Rodriguez said that further breakthroughs had been achieved, that various key technologies had been "substantially improved," that she and Baker had achieved as much as 50 percent storage capacity and 30 percent weight release of hydrogen at room temperature, and that she expected no major problems in achieving commercial status of the technology within 3 years.

In the summer of 1999, research into nanotubes for hydrogen storage got a boost from four researchers at the National University of Singapore. Writing in the July 2 issue of *Science*, P. Chen, X. Wu, J. Lin, and K. Tan claimed to have achieved hydrogen storage of about 20 and 14 percent, respectively, with lithium- and potassium-doped carbon nanotubes at moderate (200–400°C) or even room temperatures and ambient pressures. Though these numbers were not as high as those claimed by Rodriguez and Baker, they were still much higher than the numbers commonly ascribed to hydrides or other storage systems.

Baker and Rodriguez and the Singapore scientists weren't the only ones looking at nanomaterials for hydrogen storage, though. A great deal of work has been going on for several years, albeit with far less fanfare and more modest storage-capacity claims, at the National Renewable Energy Laboratory and at several other laboratories in the United States and abroad. At the spring 2000 DoE Hydrogen Program Review meeting, Michael Heben, head of the NREL effort, reported that his team had achieved storage capacities of up to 7 percent by weight at room temperatures and pressures in small samples of 1–2 milligrams, meeting or exceeding benchmark standards set by the DoE program. "In 1995," said Heben, "there was less than 1 gram of single wall nanotubes in the entire world. Today, several labs produce a gram a day." Heben's group has entered a two-year Cooperative Research and Development Agreement with Honda of America, Inc. to advance the technology.

Still, there is skepticism. Two General Motors scientists, Gary Tibbetts and Frederick Pinkerton, and a California Institute of Technology researcher, Channing Ahn, reported at the spring 2000 Canadian Hydrogen Association meeting that their respective studies of carbon nanotubes storage had persuaded them that all the aforementioned numbers were way off. All three measured storage of only about 1 weight percent or less of hydrogen at room temperatures and varying pressures, casting "serious doubts" on the claims, as Tibbetts put it.

Concern about onboard hydrogen storage was a main reason why a study by a team of four fuel cell experts for the California Air Resources Board concluded in mid 1998 that, although technology has matured to the point that fuel cell vehicles are fast becoming realistic prospects, hydrogen will not be the fuel to power them. "Hydrogen . . . is not considered a technically and economically feasible fuel for private automobiles now nor in the foreseeable future because of the difficulties and costs of storing hydrogen onboard and the very large investments that would be required to make hydrogen generally available," the four authors said.[15] But at a February 1999 conference in Vancouver, consultants who advised the Ford Motor Company on fuel cell technology called the CARB group's assessment "unduly pessimistic" and concluded that hydrogen is indeed technically feasible for onboard storage and "may very well be the least costly fuel for fuel cell vehicles, costing less than gasoline per mile driven."[16] Directed Technologies Inc. research director Sandy Thomas said the CARB study was erroneous in saying that hydrogen cannot be stored onboard a vehicle without "seriously compromising" passenger and trunk space. Thomas acknowledged that Ford had said this in one study that looked at packaging hydrogen-storage tanks into commercially existing vehicles, with no design modifications, which indeed does pose that problem. However, he added, Ford has since come up with new basic designs in which placement of the hydrogen tank is part and parcel of the package right from the start and in which the tank fits very well indeed into a car with a range of 380 miles. Apparently, Thomas said, CARB's experts "did not see these conceptual designs." "We are not suggesting that hydrogen is without problems," concluded the DTI paper. "Large hydrogen infrastructure investments will be required before fuel cell vehicles are sold; compressed hydrogen does take up more volume than gasoline or methanol tanks and the public perception of hydrogen as a dangerous fuel may impede acceptance. But in our judgement, hydrogen is much more competitive with gasoline and methanol as a fuel for fuel cell vehicles than the . . . report would have us believe."

As the new millennium was getting underway, fuel cells were receiving attention from environmental groups and from investors. Scanning the proliferating reports, releases, and web sites about fuel cells and fuel-cell-related technology and business, the initial public offerings, and the wildly gyrating stock prices of some of the better known fuel cell developers, one almost got the feeling that a new wave of feverish fuel cell stock trading, similar to the "dot-com" frenzy that had roiled the markets in recent years, was starting.

What follows is a rundown of some of the major milestones in hydrogen-fueled transportation during the last two decades, highlighted by two 1996 events: the unveiling of fuel cell vehicles by Daimler-Benz and Toyota, and the unveiling of the first close-to-commercial fuel cell passenger cars by major manufacturers.

In May 1996—3 years before DaimlerChrysler's splashy spring 1999 showing of NECAR 4 in Washington, and before Daimler-Benz merged with Chrysler—some 240 journalists from all over Europe and even some from the United States attended a media bash in the middle of Berlin's Potsdamer Platz at which the wraps were taken off Daimler-Benz's NECAR 2 fuel cell minivan. Although the two-speed transmission hiccuped because of sensor and software troubles during demonstration rides, the event was a smashing success. *The Economist*, which reported the transmission trouble, said the new vehicle's "engine ran smoothly. And this was the important point, for the NECAR 2 is claimed to be the world's first car powered by fuel cells."[17]

NECAR 2 (the dramatically slimmed-down successor to 1994's NECAR 1, a large, boxy urban delivery van whose fuel cell took up just about all of the cargo space) was a fuel cell prototype derived from Daimler-Benz's V-Class front-wheel-drive minivan. It was powered by two 25-kilowatt Ballard-type proton exchange membrane (PEM) fuel cells, one of the first payoffs of the initial $35 million cooperation and development agreement the two companies had signed in mid 1993. The white minivan, its raised roofline camouflaging compressed-gas tanks, ran on pure hydrogen. But Daimler-Benz said that 10 years or so down the line hydrogen-powered production vehicles would most likely be running on hydrogen extracted from methanol by an onboard reformer.

NECAR 2 had a range of at least 156 miles. Daimler-Benz engineers said that with prudent freeway driving it could achieve up to 250 miles at speeds of up to 69 mph between refueling stops.

In September 1997, NECAR 3 was rolled out. It was an experimental fuel cell version of the pricey five-door A-Class subcompact, which Daimler-Benz was about to unveil with conventional gasoline and diesel engines in Europe. NECAR 3's 50-kilowatt PEM fuel cell was fueled by methanol, and the early scuttlebutt was that NECAR 3 might be the basis for the company's first production fuel cell vehicle.

In October 1996, Toyota unveiled its first fuel cell vehicle, the FCEV, a PEM-equipped version of the popular RAV4, at the 13th International Electric Vehicle Symposium in October in Osaka, just about stealing the show from the almost two dozen electrics and hybrids presented by other carmakers. Toyota built two copies, at a cost of about $1 million each. With the Toyota-developed fuel cell putting out 25 kilowatts, the FCEV had a specific power output of 0.12 kilowatts per kilogram and a range of 109 miles. In addition to power from the PEM fuel cell, it used a set of nickel metal hydride batteries to recapture energy otherwise lost in braking and deceleration.

Whereas NECAR 2 relied on compressed hydrogen, the FCEV employed a titanium-based hydride. Its titanium-alloy tank stored about 2.4 percent hydrogen by weight. (The target was 3.2 percent.)

Toyota, like Daimler-Benz, was developing a parallel methanol-based version. "It's still too soon to know," Yoshio Kimura told a reporter when asked which version was likely to emerge as commercially viable. "No doubt we're moving to a hydrogen era, at which time we'd run this vehicle on hydrogen. But whether the infrastructure will be in place to coincide with our marketing plans remains to be seen. That's why we are developing both." Further evidence of that strategy was seen at the 1997 Frankfurt Auto Show when both Toyota and Daimler-Benz unveiled methanol-fueled versions of their showcase vehicles.

Kimura also agreed with Daimler-Benz that commercial fuel cell vehicles, fueled by hydrogen or methanol, were still years away: "We won't be ready to introduce it until the first part of the next century," he said.

Opel, General Motors' European subsidiary, rolled out a PEM-fuel-cell version of its then-new Zafira minivan in October 1998 at the Paris Auto

Show. The van was powered by twin 25-kilowatt fuel cells, assisted by a 20-ampere-hour, 500-watts-per-kilogram, 6.3-kilowatt-hour Ovonic metal hydride battery pack. The fuel cell Zafira was the first product of GM's new Global Alternative Propulsion Center, an international fuel cell development facility with laboratories in Michigan, in New York, and in Germany.

At the 2000 Detroit Auto Show, GM showed a fuel cell version of its Precept concept car, introduced earlier as a hybrid. The car wasn't functional, but GM vice chairman Harry Pearce said it would be by the end of the year. With a 75-kilowatt-continuous, 105-kilowatt-peak PEM fuel cell and a secret lightweight hydride, the Precept was said to have a range of 500 miles and fuel economy equivalent to 108 miles per gallon of gasoline. Pearce called it "simply a step on the long journey to a very different future that we see for the automobile."

Ford unveiled a 75-kilowatt PEM-fuel-cell version of its lightweight P2000 research vehicle at the Detroit Auto Show in January 1999. The car, roughly the size and shape of the company's standard commercial mid-size models but weighing about 40 percent less than its 3400-pound standard cousin as a result of the judicious use of aluminum and other lightweight materials, was expected to perform just about like a Taurus, with 0–60 mph acceleration of 12 seconds or so. The car wasn't quite ready for road testing: "The January auto show came a little too soon for us," a Ford engineer in charge of the fuel cell project, Ron Sims, told a reporter. "We were in the final stages of integrating all the components which we had run successfully separately." Three months later the car was up and running on Ford's test tracks, and Sims pronounced himself quite happy with it. Powered by compressed hydrogen, it had a range of only 60 miles or so—more than enough for initial development and test work in the months ahead; more range would come later. Eventually, Ford planned to replace the lightweight high-pressure compressed-hydrogen tank with an LH_2 tank that would be good for about 300 miles.

At the 2000 Detroit show, Ford showed a fuel cell prototype called Th!nk FC5 (fuel cell—fifth generation). Based on the Focus model, it was equipped with Ballard's latest 75-kilowatt fuel cell stack, the Mark 900. Unlike GM's Precept, the Th!nk FC5 was designed to run on methanol. Early copies were planned to be turned over to the California Fuel Cell Partnership[18] later that year. "This is it," said Neil Otto, the president of

Ballard's automotive division, unveiling his baby—a design frozen for commercial production—to the press in Detroit.

Also in early 2000, Korea's Hyundai announced a collaboration with International Fuel Cells to build two and maybe even four prototype fuel cell sport-utility vehicles, the first ones perhaps in time for the 2001 Detroit Auto Show.

In Europe, a joint effort by Renault and Peugeot is centered on a fuel cell under development by the Italian company De Nora. The Renault–Peugeot–De Nora program started out as a French effort partially funded by France's ADEME environmental agency. It widened in scope in early 1997 when it became the centerpiece of a new European Commission program dubbed "HYDRO-GEN." According to a paper presented at a September 1996 fuel cell conference in Chicago, its goals include the development of a 30-kilowatt PEM fuel cell that would use low-cost ($12.5/kW) electrodes and low-cost ($56/kW) membranes, which would have to last at least 2000 hours. The overall goal was a fuel cell stack costing about $250/kW initially and about $125/kW at the end of the program.

The Daimler-Benz and Toyota premieres got wide coverage in the media, in large measure because they demonstrated the growing interest of major carmakers, in the past frequently derided as too conservative when it came to breaking environmentally benign ground. However, they were by no means the first. If we discount for the moment General Motors' experimental Electrovan and Karl Kordesch's Austin (see chapter 7), that distinction belonged to small, upstart companies that saw the fuel cell's potential much earlier than the giants.

Three years before the Daimler-Benz and Toyota premieres, Energy Partners, a small company in West Palm Beach, Florida, unveiled a fuel cell vehicle that made a bit of a splash in the specialized media. Energy Partners' founder and chairman, John Perry (a millionaire and at various times a newspaper publisher, an operator of a cable TV system, and a builder of small submarines for oil exploration and for James Bond movies), had been tinkering with and supporting hydrogen, methanol, and fuel cell developments since the 1960s. Energy Partners' proof-of-concept car, a lightweight plastic-body two-seater made by Consulier Industries of West Palm Beach and converted to PEM power by Energy Partners, took to the road in October 1993. Carrying three 15-kW fuel cells in an open well, the car had a range of 60 miles in city driving and a top speed of 60 mph. Most of the development money (which totaled about $3.4 million) was put up by Perry, but $450,000 was contributed by the South Coast Air Quality Management District (the four-county smog control agency for most of the Los Angeles basin, where 15 million people breathe some of the dirtiest air in the United States).

An earlier fuel cell–battery hybrid, a converted Ford Fiesta mail-delivery car, had debuted in the summer of 1991 in Harrisburg, Pennsylvania. Pennsylvania's State Energy Office had chipped in about $60,000 in addition to what project developer Roger Billings said were millions of dollars from his own funds. Other support came from Air Products and Chemicals (Pennsylvania's preeminent producer of industrial gases) and from Exide (the largest US manufacturer of lead-acid batteries). Exide had contributed the auxiliary batteries. The car's PEM fuel cell was rated at less than 10 kW. Billings claimed that its range with the iron-titanium-magnesium hydride tank holding about 4 kilograms of hydrogen was about 220 miles.

As the new century began, early prototypes of fuel cell transit buses were being tested in North America and in Europe. Development of the internal-combustion-engine buses operating on hydrogen was proceeding too.

In 1991 the Belgian government gave formal approval to the Greenbus project. Put forward by the Belgian firm Hydrogen Systems, it called for the use of electrolytically produced hydrogen to power a converted 7.4-liter, 227-horsepower diesel city bus made by the Belgian bus maker Van Hool. The hydrogen was to be stored onboard in commercially available iron-titanium hydride storage vessels. After various modifications, the bus, renamed ZEMBUS, was launched in September 2000 in the town of Hasselt in the presence of Belgium's minister of transport. Belgium's largest transit bus company, De Lijn, expects to buy three copies and a hydrogen fueling station.

In April 1996 a bus running on liquid hydrogen was unveiled and put into regular daily operation in Erlangen, Germany. It had a 229-hp engine originally designed to burn natural gas. Another i.c.-engine bus, operating on a mixture of natural gas and hydrogen called hythane, took to the streets of Montreal in Quebec in November of 1995.

In Brazil, where air pollution is choking the life out of urban centers such as São Paulo, hydrogen buses were given some consideration in the 1990s.

Brazil's DNAEE (National Department for Water and Electricity) said in 1995 that it wanted to convert two i.c.-engine buses to hydrogen for testing. As the century drew to a close, the project was still being considered.

China issued a Request for a Proposal for a fuel-cell-powered bus in early 1998. In a separate development, the W. Alton Jones Foundation of Charlottesville, Virginia, helped to fund a multi-year project aimed at introducing fuel cell buses in Changchun (a mid-size city in northeastern China), in Shanghai, and in Ningbo (a coastal city some 175 miles south of Shanghai). The latter project died a year later, in part because Chinese supporters (such the Unirule Institute, a non-governmental agency in Beijing that supports environmentally benign technologies) decided that fuel cell technology was too expensive for a developing nation such as China.

In 1987 the first fuel cell bus project in the United States began with a competition held by the Department of Energy for the basic design of a hybrid bus to be used at Georgetown University. The competition called for a fuel cell system plus a backup battery system to help with acceleration and climbing hills and to store energy recaptured during braking. Two teams were pitted against each other. The team headed by Booz-Allen Hamilton Inc. also included the Engelhard Corporation (supplier of the fuel cell) and the Chrysler Corporation. The other team, captained by the Energy Research Corporation (supplier of the fuel cell), included the Los Alamos National Laboratory and Bus Manufacturing Systems, Inc. Four years later, when the DoE declared the first group a winner, Booz-Allen Hamilton was no longer at the helm; the New Jersey-based H Power Corporation had been named the prime contractor. The DoE's charge to H Power was to come up with three 27–30-foot-long buses within 30 months. The buses were to be powered by phosphoric acid fuel cells fueled by methanol, which was to be reformed into hydrogen with an onboard reformer (a system judged at the time to be less challenging than pure hydrogen and PEM cells). Executives from H Power and Sam Romano,[19] Georgetown University's chief investigator in the project, said at a December 1991 meeting of the Society of Automotive Engineers that, although more advanced fuel cells such as the PEM variety might be more appealing at first blush, phosphoric acid fuel cells (PAFCs) had "a lot going for them, especially on larger vehicles." For example, a PAFC's moderate waste heat could be used to run an air conditioner. And, Romano added, the PAFC was the only type of fuel cell that, at the time, had actually demonstrated acceptable performance on methanol. Methanol can be reformed into hydrogen at a relatively low temperature (200°C), Romano noted, whereas ethanol, natural gas, or propane would require about 400°C.[20] The first Georgetown bus made its debut in Washington in the spring of 1994 as part of Earth Day festivities. By that time, Engelhard had bowed out as supplier of the fuel cell system; a fuel cell manufactured by Fuji Electric was finally used. The bus itself was built by the Bus Manufacturing Corporation.

At the aforementioned 1991 SAE meeting, Ballard Power Systems provided some early glimpses of its work on Proton Exchange Membrane (PEM) fuel cells and their use as power plants for buses.[21] One of Ballard's founders, vice president Keith Prater, told the audience that Ballard had built a hydrogen-powered fuel cell weighing just under 100 pounds that generated more than 5 kilowatts when air—oxygen and nitrogen—was used as oxidizer. Prater said that the fuel cell generated twice as much power using pure oxygen, and that it would be the basis for both stationary and transportation applications. Ballard was already teamed with General Motors in a DoE program to demonstrate a methanol PEM fuel cell within 5 years, and it was also under contract to its home province of British Columbia for a hydrogen powered transit bus. Flashing a slide of a 32-foot diesel-powered transit bus on the screen, Prater said that the plan was "to put [the fuel cell] system in by the middle of next year so you can drive it around Vancouver." Unlike the Georgetown bus, Ballard's bus would not have a battery: "We wanted a bus in which there was no question that it was a fuel cell that was providing motive power, and, secondly, we wanted to explore the issue of completely powering a vehicle by fuel cells. The ultimate vehicle may well be a hybrid, but we want to explore the issue."

Ballard's first proof-of-concept bus was rolled out in January 1993. All 21 seats of the small prototype were filled by team members and well-wishers when it took its first cautious trips on company grounds. That first bus could start on a 20 percent grade, maintain 30 mph on an 8 percent grade, and accelerate from 0 to 30 mph in 20 seconds, and it had a range of 94 miles. In August 1994, the same bus made its public debut carrying real passengers during the first "green" Commonwealth Games in Vancouver.

Operating for about 6 hours every day for most of a week, it shuttled hundreds of spectators from a central dispatch point to the various athletic events.

In February 1994, Ballard announced plans for a 40-foot, 60-passenger, 275-horsepower commercial bus with a range of 250 miles and, further down the road, a 75-footer. Development of a 60-foot bus was underwritten by various Canadian government agencies and by the South Coast Air Quality Management District. Rolled out in 1995, the low-floor bus was built by New Flyer Industries of Winnipeg.

In September 1995, the Chicago Transit Authority announced that it would test three hydrogen-powered PEM-fuel-cell buses. Chicago had already tried other fuels. "For several years, the Environment Department has been experimenting with alternative fleet fuels such as biodiesel, ethanol, and natural gas," said Henry Henderson, Chicago's Environment Department Commissioner, at the press conference. "Although these offer immediate relief from air pollution, hydrogen technology could radically alter urban transportation in the next century. That's why we want to be the first ones to try it out." The buses began carrying paying passengers through Chicago's Loop in the autumn of 1997. In March 1996, Ballard signed an agreement with Vancouver's transit authority for fleet testing of three buses. When the Chicago and Vancouver test programs ended, in 2000, all concerned proclaimed their satisfaction with the results. Those programs were to be followed by trials of buses equipped with new-generation fuel cell stacks by SunLine Transit in California's Coachella Valley.

Europe's first fuel cell bus, the Eureka, made its much-delayed debut in Brussels near the end of 1994. First announced in 1988, the Eureka was an articulated 59-foot 80-passenger device whose fuel cell and other components were housed in a two-wheeled trailer almost as long as the bus itself. A hybrid with an 87-kilowatt alkaline fuel cell made by the Belgian Elenco company plus NiCad batteries from French battery maker SAFT, it had electrical traction equipment from Italy's Ansaldo and a liquid-hydrogen fuel system contributed by Air Products of the Netherlands. The bus itself came from Van Hool. In all, the partners and the member governments spent about $8 million to get the bus on the road before its demise a few months later. Elenco was forced into bankruptcy in the spring of 1995 when the shareholders refused to come up with additional cash; its assets were later bought by an Anglo-Belgian company, ZEVCO, which planned to build small airline-type hybrid fuel cell tractors but also cars and buses.

Elenco's alkaline fuel cell made a comeback and a big splash in the summer of 1998, when ZEVCO unveiled its hybrid Millennium Taxi in London. The Millennium Taxi was a traditionally styled black London cab powered by a combination of a 5-kilowatt fuel cell and batteries.

The entry in the fuel cell bus sweepstakes that probably raised the greatest expectations was Daimler-Benz's prototype NEBUS (New Electric Bus), unveiled in the summer of 1997. Another product of the partnership between Ballard and Daimler-Benz, the 12-meter low-floor city bus was powered by a ten-stack, 250-kilowatt PEM fuel cell installed in the rear; 190 kilowatts were available for traction, electrical systems, and air conditioning. Gaseous hydrogen was carried in seven roof-mounted 150-liter, 300-bar gas bottles that held about 45,000 liters of compressed hydrogen, sufficient for a range of up to 156 miles—more than enough for average daily requirements. In the spring of 2000, DaimlerChrysler announced that a commercial version, the Citaro, would be available for delivery 2 years later. In the sleeker production model, the fuel cell system is stashed on the roof to achieve a more balanced weight distribution and to extend the floor further back.

Also in the spring of 2000, the German manufacturers MAN and Neoplan rolled out two hydrogen-fueled PEM-fuel-cell buses at a "Fuel Cell Day" event in Munich. The 12-meter MAN vehicle,[22] dubbed Bayernbus I, had been in the works since 1996 and was powered by a 120-kilowatt Siemens-KWU PEM fuel cell originally designed for submarine use. Each of the four stacks was rated at 30–45 kW, for a total of 120–180 kW. Neoplan's Bayernbus II was a technologically more advanced vehicle with a unitized carbon-reinforced plastic body, regenerative braking, two hub-mounted electric motors, and an 80-kW fuel cell developed by a small startup company, Proton Motor, located south of Munich. Unlike the long-gestating MAN bus, Bayernbus II took less than a year to complete and cost only about $1.15 million in Bavarian state money—a fraction of the amount spent on Bayernbus I.

The idea of using fuel cell variants of golf carts for shopping trips began in Palm Desert, California, a wealthy and environmentally aware resort community.[23] Now hundreds of street-legal golf carts—equipped with turn signals, brake lights, etc.—are used for shopping and commuting there. Peter Lehman, director of the Schatz Energy Research Center at Humboldt State University, and Glenn Rambach, an engineer at Lawrence Livermore National Laboratory, conceived the idea in the early 1990s, when the per-kilowatt cost of fuel cell power needed for cars was still thought to be way out of economic reach. Lehman and Rambach discovered there was a sizable market for small utility vehicles of this type—about 300,000 vehicles a year.

A hydrogen-production station was a part of the Palm Desert project. Palm Desert already was building a compressed-natural-gas facility for its fleet of CNG transit buses and was evolving into a national showcase for using hydrogen in public transport. But building a hydrogen-production facility wasn't easy. "Siting the station has become a political football," Lehman told a reporter. "The process . . . has been long, arduous and politically charged," he wrote in a 1997 paper.[24]

The first hydrogen-powered golf cart, turned over by Lehman's engineers to Palm Desert officials in August 1996, was basically a standard golf cart, retaining its original 2-hp motor but equipped with a PEM fuel cell in place of the battery. It consumed about 0.29 kWh worth of hydrogen per mile—roughly the equivalent of 125 miles per gallon of gasoline, according to a Humboldt State University release. The most recent one was a Danish-built Kewet that Lehman had reconfigured to run on compressed hydrogen. More like a tiny car than a golf cart, the Kewet had a power output of 9 kW (12.6 hp). With a range of about 30 miles, this "neighborhood vehicle" was turned over to SunLine Transit in 1999 as a pool vehicle and a small test bed for that agency's ambitious plans to convert its CNG bus fleet to hydrogen power.

The Florida firm Energy Partners also saw potential in hydrogen fuel cells. It developed two concept vehicles, one called the Genesis and one the Fuel Cell Gator. The Genesis, unveiled in February 1995 in Palm Springs (near Palm Desert), was powered by a 7.5-kW PEM fuel cell and was said to be capable of carrying eight passengers or 2500 pounds of cargo. It carried hydrogen and oxygen in pressurized containers. The Fuel Cell Gator, an experimental version of a light utility vehicle built by the tractor company John Deere, operated for a while at the Palm Springs airport.

Though it seems increasingly likely that fuel cells will eventually carry the day with their inherently superior efficiency, it is by no means certain. Hydrogen internal-combustion engines have a head start. A 1995 study put the total number of hydrogen cars built at around 50, almost all of them powered by internal-combustion engines; about 20 were believed to be still running.[25]

Most notably, BMW is steadfastly sticking to internal-combustion engines in its long-running LH_2 research and development program. That firm has built four generations of LH_2 passenger cars since the late 1970s. In the summer of 1996, an elegant charcoal-gray sedan with bold lettering on its side reading BMW Wasserstoff-Antrieb (BMW Hydrogen Propulsion) conveyed Environment Minister Angela Merkel to the World Hydrogen Energy Conference in Stuttgart to give her keynote speech. (To general disappointment, Merkel said she liked hydrogen for the long term but didn't hold out much hope for it in the near term for economic reasons.)

BMW's fifth-generation LH_2 car was unveiled in the spring of 1999 at the opening of a demonstration refueling station dispensing both liquid and gaseous hydrogen (the latter for buses) at Munich's airport. The fully automatic LH_2 pump is an electro-mechanical marvel: the driver pulls up and inserts a magnetic-stripe ID card into a reader; an arm then swings out, locks onto the car's fueling port, and fills the tank with about 120 liters of hydrogen in about 90 seconds.

In 2000, as a part of BMW's "Clean Energy" program, fifteen "750 hL" sedans were deployed in several German cities. Three of these cars use a 5-kW PEM fuel cell instead of the conventional alternator and battery to generate onboard power fuel for ignition, electronics, and zero-emission air conditioning (no need to keep the engine running just to keep the car cool).

BMW essentially argues (as recently as 1998, in a paper presented at the World Hydrogen Energy Conference in Buenos Aires[26]) that it doesn't make sense to abandon the internal-combustion engine, which has undergone more than 100 years of development and refinement.

There are several reasons why fuel cells are unlikely to make it as power plants for automobiles any time soon, BMW engineer Christoph Huss asserted in his paper. For one thing, he explained, an electric traction motor for a fuel cell vehicle costs as much as a normal internal-combustion engine.

And the fuel cell, costing between $3000 and $5600 per kilowatt, would drive up the cost of the entire power train by a factor of 100 relative to an internal-combustion engine.[27] Furthermore, said Huss, fuel cell engines are likely to remain three times as heavy as internal-combustion engines for the foreseeable future; nobody has any experience with mass producing them; though more efficient than i.c. engines when running on hydrogen, they are less efficient than i.c. engines when operating on methanol; requirements for platinum as catalyst material are still high (Huss claimed 4 grams per kilowatt); and fuel cells are not suitable for dual-fuel operation—an approach favored by BMW—and therefore could not be widely deployed until a widespread alternative fuel infrastructure had been set up.[28] Thus, declared Huss, "the fuel cell will not be suitable for driving a vehicle in the foreseeable future."

BMW is not alone. In Japan, the Musashi Institute of Technology, long in the international forefront of developing LH_2 power for automobiles, has built several generations of LH_2-powered i.c.-engine cars and trucks, including a racy Nissan sport-coupe conversion and a produce-delivery truck. In an exhibit on the sidelines of the Kyoto Global Climate Change conference in December 1997, Musashi displayed an LH_2-fueled Nissan station wagon, redubbed Musashi 10, that was said to have a range of 300 kilometers. And Ford, a partner in the transatlantic fuel cell alliance with DaimlerChrysler and Ballard Power Systems, announced in mid 1999 that it had begun looking at hydrogen-powered i.c. engines again and was planning to install them, with both gaseous-hydrogen and liquid-hydrogen feeds, in an experimental lightweight P2000. Furthermore, the Department of Energy and most of the international carmakers are working feverishly to improve i.c. engines using conventional carbonaceous fuels so as to bring their polluting emissions down to extremely low levels and their efficiency and mileage to unheard-of high levels. The DoE, through its Partnership for a New Generation of Vehicles (PNGV) program, is pushing development of a small, diesel-fueled, compression-ignition, direct-injection engine (a high-tech diesel, as it were, although some program managers bristle at that description) as a relatively low-cost alternative-fuel engine with what PNGV managers say are "demonstrated manufacturability and affordability"—areas in which fuel cells have a way to go. At some point, there may well be a new meeting and mating of high-tech, super-efficient, advanced internal-combustion engines with advanced technology for the combustion, production, and storage of hydrogen.

Daimler-Benz's hydrogen i.c. vehicles have racked up half a million miles, proving, said the company in an early paper, that "hydrogen is suitable as a fuel" and that "the safety standard achieved can be compared to that of conventional vehicles." Daimler-Benz began investigating hydrogen-powered vehicles in earnest back in 1973, beginning with a hydrogen version of an early minivan that was eventually introduced at the 1975 Frankfurt Auto Show. To carry the fuel, that van used a titanium-iron hydride, mainly because the Daimler-Benz researchers were in a hurry to demonstrate the general feasibility of hydride storage and hydrogen as an automobile fuel to a wide public. "We completed the engine in about two weeks," recalled Helmut Buchner, who ran the hydrogen-hydride program at the time.

Except for various test gauges mounted on the dashboard, that first Daimler-Benz hydrogen bus was indistinguishable from the production model. One had to look very hard to notice anything different on the outside. The gas filler cap was replaced by a small hydrogen intake valve mounted on the right side underneath the body, just behind the front wheel. And the exhaust felt moistly lukewarm, almost like the steam from a teakettle that's just warming up. This was the water exhaust vapor, which was partially recycled through the engine to prevent backfiring.

In the early 1970s, when I was a correspondent for McGraw-Hill World News and *Business Week* in Bonn, Daimler-Benz invited me to take a ride in that van around their Stuttgart test track. Nothing seemed out of the ordinary. Engine noise appeared to be normal, and the van went through the banked turns at a steady 60 miles per hour. At one point, the driver stopped the bus on a 10° incline and then drove off smoothly in first gear. The 17-gallon tank carrying 440 pounds of titanium-iron hydride was stored under a bench seat. To release the hydrogen, heat from the engine radiator was cycled via a heat exchanger through the hydride tank. Rather than a conventional carburetor, there was a Gasmischer (gas mixer) to blend air and hydrogen; the mix was then sucked through the manifold into the engine for normal combustion.

The largest early attempt to gauge the potential of automotive hydrogen technology was a four-year fleet test of ten hydrogen-powered Daimler-

Benz vehicles in what was then West Berlin. Five dual-fuel (hydrogen and gasoline) station wagons and five full-size hydrogen-only vans were turned over, after intensive training, to drivers from a Berlin emergency medical service, the city government's car pool, an ambulance service, the German Red Cross, a study group for efficient energy utilization, the Berlin gas utility GASAG, and a Jewish Community organization, according to a 1990 summary report.[29] Begun in May 1984, the test ended in March 1988, after racking up more than 160,000 miles. Another 238,000 miles were accumulated in Stuttgart by prototype vehicles before the start of the actual fleet tests.

The engines of all these vehicles operated in the "external mixture formation"[30] mode, with hydrogen blown into the intake manifold. This is technically relatively simple, requiring only low-pressure hydrogen gas released normally by hydrides. But it also results in fairly low power and irregular combustion—backfiring. The solution used at the time—injecting water into the suction manifold in the upper load range—was an acceptable compromise in terms of fuel consumption and power output; it also cooled some critical components.

The hydrogen was extracted from "town gas" (which contained roughly 50 percent hydrogen) via pressure-swing adsorption. The extraction was done in a small, specially constructed pressure-swing adsorption plant, complete with 200-bar pressure storage tanks near the actual refueling station, designed and built by Bergbau-Forschung GmbH. The fueling station was operated by ARAL, Germany's largest chain of retail gas stations. Standards for fuel quality were tightened when it became evident that the relatively low purity (initially, 99.995 percent) created contamination problems and reduced the hydride's storage capacity. Hydride contamination led to a steady reduction in the range of the vehicles, requiring removal of the storage units after 2 years and reactivation of the hydride material. Impurities in the injection water increased engine corrosion and caused malfunctions in the mixture formation systems. In four of the five vans, corrosion was so bad that the engines had to be replaced. Flexible stainless steel tubing in the hydrogen piping system also was corroded by de-icing salt used in Berlin's streets. Mechanical damage was caused to heat exchangers and exhaust systems by the reduced road clearance, which was due to the extra weight of the hydride systems, according to the final report.

Nevertheless, the four-year test was judged an overall success: "On the whole, experience was positive and this was in large measure due to the fact that the vehicles were well received by the operators and the maintenance and service personnel. It was possible to operate the vehicles practically all year round without limitations and their reliability was very satisfactory on the whole. . . . Failures which occurred were to be expected in view of the experimental nature of such a complex and innovative overall system. No basic doubts arose about the project at any time during testing."[31]

BMW, Daimler-Benz's principal German competitor, began to explore hydrogen technology in 1979. That year, BMW's first liquid-hydrogen car was put together by Walter Peschka from parts and components provided by BMW. Peschka, a researcher at the German aerospace research agency DFVLR in Stuttgart,[32] and a colleague, Constantin Carpetis, devised the world's first LH$_2$ fuel tank specifically designed for passenger cars. A flattened stainless steel sphere with two separate shells (one inside the other, with a vacuum in between), it was capable of holding about 29 gallons of LH$_2$ at a low operating pressure of 4.5 bars. Peschka and Carpetis also developed a semi-automatic LH$_2$ service-station pump, which they said was so simple to operate that it might be suitable for a self-service station. As described at the 1978 World Hydrogen Energy Conference in Zurich, the device could pump between 120 and 150 liters of liquid hydrogen in about 5 minutes.

In 1989, BMW built the world's first test stand for hydrogen-powered internal-combustion engines. BMW also converted one of its limited-production V-12 engines to liquid hydrogen. The engine was displayed that autumn (more or less "for the fun of it," said one BMW hydrogen researcher) at the Frankfurt Auto Show. Though still "a long way from production," the engine was "a good indication of how BMW sees the future," *Road & Track* magazine said.

In June 1994, BMW displayed all four generations of its liquid-hydrogen-fueled internal-combustion-engine-powered cars at a one-day engineering symposium to which it had invited some 300 journalists from all over Europe. Also on display were three dual-fuel (compressed natural gas and gasoline) demonstrator sedans which BMW was launching. BMW clearly regarded compressed natural gas and liquefied natural gas as logical steps toward the use of liquid hydrogen. BMW believed that California's looming ZEV standards could be met by an internal-combustion engine only if hydrogen were to be the fuel. But solar hydrogen—the only type of hydrogen that BMW's engineers regarded as environmentally acceptable— cost about US $11.25 per liter of gasoline equivalent.

Still, the conference opened many eyes to the practicability and safety of LH$_2$ technology for cars. Wolfgang Strobl, BMW's principal hydrogen investigator, showed dramatic slides illustrating liquid hydrogen's safety: Double-walled tanks filled with the super-cold fuel and having all safety valves blocked were cooked over high heat, violently shaken in long-term vibration tests, and rammed with a massive pole (simulating the impact of a highway crash). In a fire test, hydrogen began escaping from safety valves after sitting about 10 minutes in an open fire; it burned without any other visible effect on the tank. The pole's impact produced a leak through which hydrogen escaped slowly, but, said Strobl, the tank did not explode. In other instances, though, with extreme pressure buildups, some tanks designed to then-current safety standards did burst—something Strobl acknowledged would have to be addressed.

The third major carmaker to jump into the hydrogen pool was Mazda, which began to investigate hydrogen as an automotive fuel around 1986. That effort culminated in the futuristic HR-X (Hydrogen Rotary Experimental) concept car unveiled at the 1991 Tokyo Motor Show—the first of eight hydrogen-powered Mazdas (two copies each of four different models). Visually and technically, the HR-X was the most radical of the lot. This hydrogen hybrid was powered by a 1-liter-displacement Wankel-type rotary engine mounted in the center of the chassis. The engine produced 100 horsepower running on hydrogen—outstanding performance for an engine of its size at the time. Mazda was the last remaining adherent of Wankel technology; other carmakers had experimented with Wankel engines but had dropped them because of sealing problems and other technical problems. Undaunted, Mazda said that the Wankel engine, with its separate intake and combustion chambers, was uniquely suited among internal-combustion engines to burn hydrogen, being free of the kind of backfiring and ignition problems that plagued other hydrogen internal-combustion engines at the time.

The HR-X made its US debut at the New York Auto Show in April 1992. Takanori Minami, general manager of Mazda's technical research center in

Hiroshima and the man mainly responsible for the car's development, told me that Mazda was tackling several "convenience areas" to make the HR-X and hydrogen power a viable solution, "including establishing and supporting the infrastructure necessary for distributing hydrogen." "We really want," he said, "to . . . show that hydrogen can be utilized in smaller vehicles like the HR-X. We want to appeal to the people."

In 1993, Mazda introduced the HR-X2 at the Tokyo Motor Show. The HR-X2, with a front-mounted rotary engine driving the front wheels, looked much more conventional than the original HR-X and offered much better performance. It had a better range (143 miles, versus 120 for the HR-X) and a higher top speed (90 mph, versus 81).

More or less simultaneously, Mazda developed a hydrogen-powered version of its popular Miata sports car, the main difference being that the standard 1.6-liter four-cylinder engine was replaced by a 1.3-liter twin-rotor engine. The main rationale for this conversion project was that it provided a basis for comparison with three battery-powered Miatas that had been built earlier and were being operated by a Hiroshima utility. For sports car aficionados, the killing drawbacks would have been the 770 pounds added by the hydride storage system, the reduced performance, and the reduced driving range (only about half of that of the standard Miata).

In 1994, Mazda converted two station wagons to hydrogen, again using the 1.3-liter twin-chamber rotary engine. The main purpose of this effort was to evaluate durability, refueling characteristics, and general usability under real-world driving conditions over a longer term. The cars were used every day in the motor pool of Nippon Steel's Hirohata Works, which generates hydrogen as a by-product. Together these two cars covered 12,500 miles in a two-year test of mostly city driving, Kazukiyo Okano of the Engineering Advancement Association of Japan reported at the 1997 National Hydrogen Association meeting in Alexandria, Virginia. "These cars," said Okano, "proved hydrogen vehicles are safe and can be used in the city without any problems."

Mazda made the move to fuel cell power in December 1997 when it unveiled a hybrid fuel cell station wagon, the Demio FCEV, on the sidelines of the Kyoto Global Climate conclave. That vehicle had a relatively low-powered 20-kilowatt PEM fuel cell that, in a novel twist, was augmented by a 20-kW ultracapacitor instead of the usual batteries.

Walter Peschka's involvement with BMW[33] and hydrogen was an out-growth of a project, begun in 1979, in which Germany's DFVLR aerospace agency teamed up with the Los Alamos Scientific Laboratory and the New Mexico Energy Institute to convert a Buick Century sedan to LH_2 power. As Peschka recounts in his book *Liquid Hydrogen—Fuel of the Future* (Springer-Verlag, 1992), Los Alamos was in charge of the overall coordination, including management, engine conversion, tests, and methodology. DFVLR provided the aluminum LH_2 tank and an electronically controlled semi-automatic refueling station. The car, powered by a turbocharged 3.9-liter 6-cylinder engine, was modified by the Billings Energy Corporation of Independence, Missouri. In his book, Peschka described the car's acceleration as "not completely satisfactory" at Los Alamos's altitude (7300 feet): "The turbocharger which normally compressed the fuel-air mixture in the gasoline version and (which) was not designed for the hydrogen-air mixture could only partially eliminate the loss of power output due to altitude and external mixture formation." The LH_2-powered Buick took to the road in May 1980. Though driven mostly around Los Alamos, it was also demonstrated elsewhere in the United States.

As early as 1973, scientists at Los Alamos had converted a sturdy half-ton Dodge pickup truck to LH_2 power. That truck had a spherical 190-liter aluminum storage dewar (a Thermos-bottle-like double-walled container), designed and fabricated by the Minnesota Valley Mining and Engineering Company, in its bed, right up against the forward wall. The fuel powered a 5.2-liter V-8 via external mixture formation, the carburetor having been replaced by an IMPCO gas mixer originally designed for liquefied petroleum gas and propane. Exhaust gas was recirculated to avoid uncontrolled pre-ignition, knocking, and backfiring. The vessel's boiloff rate was less than 1 percent per day, but in actual practice—during filling and discharge of hydrogen, and during measurements—the rate was more like 3 percent. Since the vehicle was apparently a sort of enthusiasts' project with no real research funding and sponsorship, it underwent only limited testing, accumulating only about 300 miles over roads in and around Los Alamos.

Of the 50 or so hydrogen vehicles thought to have been built so far, a sizable fraction were powered by liquid hydrogen. Peschka's book documents 15 of them that were constructed between 1970 and 1988. Among these are a Winnebago motor home and a Chevrolet Monte Carlo converted by the Billings Energy Corporation in the early 1970s; the aforementioned BMWs; and, beginning in 1975, the first in a series of passenger cars built by the Musashi Institute of Technology.

The Musashi 9, a refrigerated truck, was shown in Yokohama at a New Energy conference in 1993. A conversion of a 4-ton tilt-cab Hino, it made ingenious use of liquid hydrogen's cryogenic properties not only to power the 6-liter, 160-horsepower diesel engine but also to keep produce, fish, or other perishables fresh during transport. It was visually distinguished from its conventional cousins by a 400-liter cylindrical LH_2 storage tank mounted vertically on the left side between the cab and the cargo box. Kimitika Yamane, an assistant professor at Musashi, explained at the Yokohama conference that 400 liters of LH_2 is roughly equal in energy content to the 100 liters of diesel fuel the normal version would carry onboard—enough for the Tokyo-Osaka run of about 375 miles. To accommodate the tank, the cargo container had to be shortened about 47 inches. The location was determined primarily by safety considerations; behind the cab and ahead of the storage area, it was in the spot regarded as least likely to be damaged in a crash. The fuel tank was connected to a refrigerant reservoir and then to the cargo ice box, which had to be kept at a constant temperature plus or minus 5°C; that required about 17 horsepower of the engine's total power output.

A great deal of the early hydrogen development work was performed by enthusiastic individuals and hydrogen advocates in the 1960s and the 1970s. Amateurs as well as professionals like Walter Peschka were not only beginning to think about clean fuels; they were also beginning to construct cars and hardware, picking up where Rudolf Erren and his contemporaries had left off three decades earlier.

Back in 1966, for example—3 years before Buckminster Fuller was telling college audiences how to use the world's energy "current account" instead of robbing the global energy "savings account"—a high school student piloted a wildly backfiring Model A Ford truck through the quiet residential streets of Provo, Utah. With the help of his father, a teacher, and some friends, 16-year-old Roger Billings had spent some 3 months and 800 working hours not only restoring the Model A but making technological history.

That truck, which delivered little power but "an amazing repertoire of noises," according to Billings, was apparently the first internal-combustion-engine car in the United States fueled by hydrogen. Billings had started tinkering with hydrogen as a fuel 2 years earlier. "Those were the proudest days of my life," he recalled later, "driving that truck on a fuel they said would never work." In the 1970s and the 1980s, Billings was a prominent member of the small international hydrogen brotherhood.

Interspersed between Billings's early effort and the Daimler-Benz prototype of 1975 was a great deal of research, much of it done by university scientists. "It is not the internal-combustion engine that pollutes our air, but its present fuels," declared the late Kurt Weil, then a professor emeritus at Stevens Institute of Technology, at the landmark 1972 Energy Conversion Conference in San Diego. The German-born Weil had experience with hydrogen going back to the 1930s, when he and Rudolf Erren suggested a scheme to use the excess capacity of Germany's electric power grid to produce hydrogen through electrolysis. Proposing a similar plan for the United States, Weil told his fellow researchers:

The central element of such a system is the hydrogen-burning internal-combustion engine which already exists in practical and proven models. Its multi-fuel and mixed-fuel version offers not only complete adaptation of all existing internal-combustion engines (about 150 million engines in the USA alone) but it also allows complete flexibility in
• phasing out hydrocarbon fuels when, to what degree, and at which rate desired, and geographic adaptation to various available fuels and mixtures;
• reducing or eliminating air pollution from internal-combustion engines, including carbon dioxide, nitric oxides and particulate matter to any degree, at any time of danger and at any region.

Perhaps most representative of the hydrogen enthusiasts who were stimulated by the emerging environmental movement and wanted to exploit hydrogen's unique properties were four members of the Perris Smogless Automobile Association, a group of residents of Perris, a small California town south of Riverside. The four—a civil engineer, a newspaper publisher, and two aerospace engineers—shared a commitment to ridding the environment of what they considered its worst enemy: automotive pollution. Patrick Lee Underwood, one of the aerospace engineers, had worked for the aerospace firm Lockheed in various research programs. He had thought about hydrogen as an aviation fuel in the late 1960s after hearing about an experimental hydrogen-fueled B-57 bomber that the US National Advisory Committee for Aeronautics had test flown successfully in 1957. The First Annual Report of the Perris Smogless Automobile Association, published in 1971, stated that the idea of "using hydrogen and oxygen in a standard reciprocating automobile engine had occurred to Mr. Underwood several years ago but no constructive effort had been expended; indeed, the idea was not very well developed, generally because it was thought to be under development as a natural course by the aerospace industry."

The actual project had gotten underway in December 1969, when Underwood had challenged an editorial, written by Dwight Minnich for his paper, the *Perris Progress*, in which Minnich had asserted that there was no solution to the emission problem of the internal-combustion engine. A month later, Minnich, Underwood, and Fredric Nardecchia got together in Minnich's office to sketch out the outline of a hydrogen test program. A short while later, they were joined by Paul Dieges.

The Perris experimenters' first test vehicle was a 1950 Studebaker bought for $10 from a junk dealer. Several attempts with various hydrogen-oxygen and hydrogen-air mixtures failed, but eventually the men succeeded in starting the engine after feeding compressed gaseous hydrogen into the manifold intake. The tests were run by "remote control from behind a large rock." After adjusting gas pressures and carrying out some other tests, the group established that the engine could be made to run quietly, start easily, and respond rapidly to control changes, that it did not knock or display "any other disturbing sounds," but that it used an unexpectedly large amount of hydrogen. The Studebaker project was abandoned after the engine failed to restart after a test in which it had been grossly over-revved on too much hydrogen.

In their next project, the Perris experimenters modified a 1930 Model A Ford pickup truck for road trials, again burning compressed hydrogen and compressed, almost pure oxygen. The arrangement also included a condenser to cool the exhaust vapor back to water, with excess hydrogen to be recirculated to the combustion process. The truck, licensed as an antique, made its first run in February 1970, covering about a mile. Later it made several trips through and around Perris, plus some demonstration runs for officials of the California Air Resources Board. In a paper presented at the 1971 Intersociety Energy Conversion Engineering Conference, Underwood and Dieges reported that the truck ran fine. The exhaust water vapor was continuously accumulated and drained off; "the water was found to be potable but somewhat oily and rusty tasting."

Also in 1970, the Perris experimenters converted a newer Ford pickup truck, an F250 with a 6-cylinder engine, to run on liquid hydrogen and liquid oxygen. "The whole idea was to have absolutely no impact on the atmosphere," Dieges told me in a March 1999 telephone conversation.

Next the Perris group planned to convert a donated 1952 Cadillac to hydrogen, but those plans were aborted when a coast-to-coast Clean Air Car Race was announced. This event, sponsored by Caltech and MIT, would be open to all types of nonpolluting or low-polluting cars. The Perris Smogless Automobile Association had obtained a grant from General Motors to help prepare a pickup truck to operate on LH_2 and liquid oxygen for the race, but unfortunately the project was stopped by a variety of logistical and managerial problems. For example, they could not get enough small vacuum-jacketed dewars to carry enough LH_2 to propel the car from one major refueling point to the next. (Linde, an American hydrogen manufacturer, had offered to supply the required LH_2; the problem was to bridge the large gaps between the various distributorships.) In the course of race preparations, the Perris team developed an "oxyburetor," analogous to a carburetor, which drew in a proportionally metered amount of oxygen to be mixed with hydrogen. The truck eventually ran in preliminary trials, reaching a top speed of about 40 mph, but its ability to climb hills was "poor" due to an insufficient oxygen flow rate limited by inherent restraints of the oxygen cylinders and the flow-rate regulators.

The most poignant and revealing part of the Perris group's "first annual report" (there was no second[34]) was the balance sheet. Expenditures totaled $8038 for a whole year's work, with major individual contributions of $2465.68 by Underwood, $1828.68 from Minnich, and close to $400 each from Dieges and Nardecchia. General Motors chipped in $2180, the California Society of Professional Engineers $300. The biggest single expenditures were $1429.01 for patent fees, $1142.46 in travel costs for the Clean Air Car Race, and $877.72 for exotic parts for the noncompeting Ford pickup truck. The hydrogen cost $241.84. (Linde contributed some hydrogen free of charge.) The oxygen bill ran to $50.20.

The language of the "first annual report" shows humility as well as a sense of mission and excitement. In the preface, the four principals described themselves as "average business and professional men with average means and the usual family and personal obligations." They continued: "We have, of necessity, pursued our regular livelihoods during the course of this effort. These remarkable results, which have vastly exceeded even our own expectations, have thus far been elicited by a commitment of only somewhat higher order than we might have otherwise directed toward civic, social, church or hobby activities, plus our vacations which all of us devoted to this project." Their sense of outrage at the ravages upon the environment is clear in the report's opening words:

Future historians, if there are any, may well record that the private automobile and not the nuclear bomb was the most disastrous invention of a society so obsessed with technology that it never recognized the failures of engineering run rampant until too late. For the automobile-freeway system is surely the most inefficient, dangerous, costly, and environmentally damaging transportation ever conceived; about the only real plus is flexibility, and the actual popularizing features are frighteningly Freudian. . . . survival demands that the air pollution caused by the automobile be eliminated almost immediately.

Strong words from three engineers and a small-town newspaper publisher!

As time went on, other organizations became more interested in hydrogen's potential as fuel and were eager to tout its virtues. An early example of such promotion was a film clip produced by the Jet Propulsion Laboratory in the mid 1970s. First, the camera focused on the exhaust pipe of an idling car emitting white vapors. Next, a man held a drinking glass against the exhaust tube, allowing the vapors to cool and condense into a colorless liquid. In the startling final shot, the man lifted the glass to his lips and drank what was almost pure water—a stunt that by now has become an almost standard photo opportunity for hydrogen demonstrations to show off its environmental cleanliness. (Mayor Richard Daley of Chicago had to gulp the waters from an idling Ballard PEM-fuel-cell bus in September 1995 when the city started its fleet test; ABC TV correspondent Ned Potter did the same in a *World News Tonight* feature on the promise of fuel cells in April 1997.)

Hydrogen power won one of its early public validations as well as public attention in the 1972 Urban Vehicle Design Competition. Roger Billings, by then a graduate student in chemistry at Brigham Young University, entered and won the anti-pollution category with a hydrogen-fueled Volkswagen that far exceeded the existing federal clean air standards.

The runner-up in the anti-pollution category, and the winner of the overall design category, was a team of students from the University of California at Los Angeles that had entered a hydrogen-burning American Motors Gremlin. The captain of the UCLA team was a student named Frank Lynch. Right after the results were announced, Billings and Lynch decided to team up. The result was the Billings Energy Corporation of Provo, Utah, established to promote and eventually make money off the hydrogen-fueled automobile. The Billings Energy Corporation won recognition during the 1970s and the 1980s for converting various vehicles—passenger cars, a motor home, a small bus—to hydrogen.

Around 1979, Frank Lynch formed his own company, Hydrogen Consultants, Inc., in Littleton, Colorado. Renamed Hydrogen Components Inc. in early 1997, the firm is now internationally recognized as one of the most savvy constructors of equipment for hydrogen-powered vehicles. For several years in the early 1990s Lynch was a member of the US Department of Energy's Hydrogen Technology Advisory Panel, a committee charged with advising the Secretary of Energy.

During the 1960s and the 1970s, there were at least a dozen efforts by academic researchers or private companies to employ hydrogen, either alone or as a supplement to regular gasoline, in the quest for fewer unburned hydrocarbons, carbon monoxide, and NOx (nitrogen oxides) emissions. Some examples:

• In 1970, two University of Oklahoma scientists, Roger Schoeppel and Richard Murray, under contract to the US Environmental Protection Agency, adapted a 3.5-horsepower four-stroke engine to run on injected hydrogen. The modifications included installing a second camshaft system to operate the hydrogen-injection valve and installing a new water cooling system to improve the cooling characteristics of the cast iron cylinders. Tests were encouraging, Murray and Schoeppel said in a paper. Fuel flow was only one-third that of gasoline, torque was fairly constant over a wide speed range, and power could be boosted beyond the manufacturer's rating simply by injecting more hydrogen.

• Two University of Miami researchers, Michael Swain and Robert Adt Jr., converted a Toyota station wagon, lent by a Miami Toyota dealer, to run on hydrogen. Swain and Adt devised a new fuel system, dubbed HIT

(Hydrogen Induction Technique), that took advantage of hydrogen's high flame speeds and wide range of combustible fuel-to-air ratios. Normally, stepping on the accelerator pedal increases the amounts of both fuel and air going into the engine, but Swain and Adt modified the fuel system so that it would change the fuel/air ratio. For more power, more hydrogen but the same amount of air—a richer mixture—was injected, similar to the diesel principle. Swain and Adt reported fuel consumption very roughly equivalent to 42 miles per gallon of gasoline—a 50 percent improvement in brake thermal efficiency over the standard gasoline version of the Toyota wagon.

• Reforming gasoline into hydrogen was described at the 1972 San Diego energy conversion conference by Harold Sorensen of the International Materials Corporation. In his Boston Reformed Fuel Car project, Sorensen proposed to burn only hydrogen in the engine, with CO_2 and water vapor as the only emissions.

• In 1973, Siemens, a German maker of electrical equipment, announced a "crack carburetor" that was to use a catalytic process to break down gasoline into methane, hydrogen, and carbon monoxide. The project was dropped two years later because Siemens couldn't find any takers.

• In 1974, the Jet Propulsion Laboratory proposed combusting hydrogen together with gasoline at "ultra-lean" conditions. Hydrogen was generated aboard the car from gasoline through the use of hot air (1500–2000°F). Unlike the Boston car, in which all gasoline would have been converted to hydrogen, only small amounts would be produced. Hydrogen would extend the flammability limits of the fuel-air mixture downward, requiring less fuel with the same amount of air. The car did turn out to be more efficient and to produce less NOx, but at a power loss, and other emissions were still high.

• In Detroit, two General Motors engineers also found that by injecting hydrogen into unleaded gasoline the "lean limit" could be reduced further and NOx emissions could be cut drastically, but at a cost. R. F. Stebar and F. B. Parks said at a spring 1974 SAE conference that adding about 10 percent hydrogen lowered the lean limit by almost ⅓ and reduced NOx emissions to ¹⁄₇₅. Carbon monoxide levels stayed about the same, but hydrocarbon emissions went up dramatically. The engine produced only about two-thirds of its normal power.

Hydrogen work also was going on in the Soviet Union. At the 1976 Miami Beach hydrogen conference, Soviet scientists said in hallway conversations that they had experimentally converted a couple of Soviet-built Fiats to hydrogen use. Two years later, a Moskvich sedan was also modified to run on gaseous hydrogen, according to a report in the March 1978 issue of *Eastwest Markets*, a business publication covering trade with the communist bloc. The story said that "the vehicle worked well on a test track and emitted pure water vapor as exhaust." A November 1978 article in the newspaper *Socialist Industry* said that Soviet scientists had successfully tested a Volga automobile on a mixture of gasoline and hydride-stored hydrogen.

In 1988, a World Hydrogen Energy Conference was held in Moscow. It turned out to be an eye opener for many Western visitors. Soviet scientists presented 75 papers, about half of the total. Hydrogen R&D "appears to be a serious effort in the Soviet Union," observed Alexander Stuart, president of Canada's Electrolyser Corporation and one of the key figures of the international hydrogen energy scene. "It seems to enjoy a high level of attention."

Some Western scientists who visited research institutes as part of the program noted that there was quite a bit of outdated equipment side by side with modern instruments. By and large they came away impressed with the breadth and scope of the work, the thoroughness of research efforts, and the number of scientific workers in the field. One Soviet researcher estimated that some 500 scientists were active in hydrogen energy-related work. The roster of the conference listed about 15 institutes spread across the Soviet Union.

One highlight of the conference was the keynote speech, in which the aircraft designer Alexei Tupolev detailed the efforts that had led to the maiden flight of a partially hydrogen-powered TU-155 commercial jet earlier that year. Another highlight was the LH$_2$-powered RAF (Riga Automobile Factory) 2203 minivan that was displayed outside Moscow's International Trade Center on the last day of the conference. And there were papers and posters describing such projects as the conversion of a Lada (a 1970s Fiat design licensed to the Soviet Union) to run on hydrogen, the installation of a fuel cell in another RAF van, the conversion of several other RAF vans to gasoline-hydrogen operation, and plans for the installation of a 40-kilowatt hydrogen-air fuel cell in a Hungarian Ikarus.

The 1988 conference turned out to be the last hurrah for large-scale hydrogen work in the pre-collapse USSR or in Russia. Though Russian scientists continued to present papers on various hydrogen projects at international meetings, there was little evidence that any real efforts were still underway. For the remainder of the twentieth century, hydrogen R&D's center of gravity had shifted to the West.

A German self-propelled rail car converted by Rudolf Erren to run on hydrogen. *(photograph provided to the author by Erren in the late 1970s)*

Bill Powers, Ford Motor Company's Vice President of Research, discusses the fuel-cell-powered P2000 during a press conference held on the occasion of the opening of a hydrogen fueling station at Ford's Dearborn research center, August 1999. *(courtesy of Ford Motor Company)*

Europe's first hydrogen fueling station, in Hamburg, shortly after the ribbon-cutting ceremony in January 1999. The station is named after the German hydrogen pioneer Ludwig Bölkow. *(courtesy of Hamburg Hydrogen Agency)*

Journalists awaiting their turn to test drive DaimlerChrysler's experimental Necar 4 fuel cell car during the vehicle's introduction, Washington, March 1999.

Toyota's FCEV. *(courtesy of Toyota)*

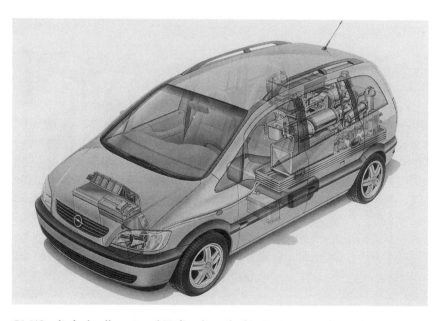

GM/Opel's fuel-cell-powered Zafira, launched in Paris in October 1998. *(courtesy of Opel)*

Ford's Th!nk FC5 fuel cell prototype at the 2000 Detroit Auto Show.

The Proton Motor/Neoplan Bavaria II fuel cell bus, shown during a "Fuel Cell Day" event in Munich in May 2000. *(courtesy of L-B-Systemtechnik)*

Peter Lehman of Humboldt State University in his hydrogen-fuel-cell-powered Kewet at the Palm Springs Convention Center during the November 1998 Fuel Cell Seminar.

The hydrogen fueling station at Munich's airport. *(courtesy of BMW of North America)*

Daimler-Benz's first hydrogen-powered minivan, launched at the 1975 Frankfurt Auto Show. *(courtesy of Daimler- Benz)*

Ned Baker, a member of the 1972 UCLA Urban Design Vehicle Competition team, graphically demonstrates the cleanliness of an engine fueled by hydrogen. The almost pure water he is drinking condensed out from the steam coming out of the tail pipe. *(prints from 16-mm film clip produced by UCLA)*

Karl Kordesch's hydrogen-fuel-cell-powered 1961 Austin A-40. *(courtesy of Prof. Kordesch)*

The experimental Allis-Chalmers fuel cell golf cart of the early 1960s. *(courtesy of Jay Laskin)*

The Allis-Chalmers fuel cell tractor of the early 1960s, perhaps the first fuel cell vehicle ever. *(courtesy of Jay Laskin)*

General Motors' fuel cell Electrovan. *(courtesy of GM Archives)*

The Tupolev 154 passenger jet, with one engine converted to run on hydrogen. *(courtesy of DaimlerChrysler Aerospace Airbus)*

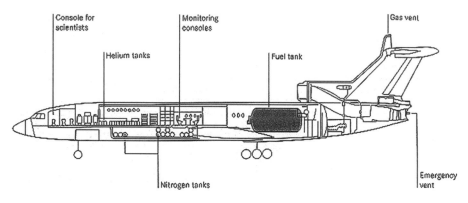

A schematic of the experimental hydrogen-fueled TU-154. The fuel tank in the rear fuselage was added to carry liquid hydrogen. *(courtesy of DaimlerChrysler Aerospace Airbus)*

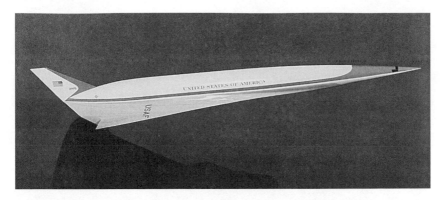

A model of the National Aerospace Plane. *(courtesy of US Air Force)*

A rendering of the "Cryoplane." *(courtesy of DaimlerChrysler Aerospace Airbus)*

Olof Tegström's hydrogen-heated house and hydrogen-fueled Saab. *(courtesy of Olof Tegström)*

Walt Pyle's solar hydrogen house. *(courtesy of Walt Pyle)*

The solar hydrogen house built by the Fraunhofer Institute for Solar Energy Systems in Freiburg, Germany. The hydrogen tank is in right foreground. *(courtesy of Fraunhofer Institute for Solar Energy Systems)*

Fuel Cells: Mr. Grove's Lovely Technology

At first glance, the small black sedan tootling around the Cleveland suburb of Lakewood in the early 1970s seemed just another foreign economy car. Only on a closer look did its unusual features become apparent. Mounted conspicuously crosswise on the roof were six scuba-type gas tanks. And in front of them, up over the windshield, was a sign reading "compressed gas—flammable." In fact, this 1961 two-door Austin A-40 was the harbinger of a revolution in automotive technology that was not to burst onto the international scene until a quarter of a century later. Constructed in 1970 by an Austrian scientist who had been spirited out of his native country by the US Army, it was the world's first practical fuel cell car.

Karl Kordesch left Vienna in 1953, ostensibly on a vacation with his family and his two small children. After a somewhat surreptitious passage through the Soviet-controlled part of Lower Austria, he and his family eventually turned up in New Jersey with the US Signal Corps under the auspices of Project Paper Clip. Two years later he joined Union Carbide's research laboratory in Parma, Ohio, as a staff scientist working on batteries. At Union Carbide he eventually became a Corporate Research Fellow; this meant, he recalled, "complete freedom, including worldwide travel, and getting more money without the need to become a manager." Kordesch produced many patents and published lots of scientific papers.

The air-hydrogen alkaline fuel cell of Kordesch's Austin produced a tepid 6 kilowatts in conjunction with a bank of ordinary 12-volt lead-acid batteries (for acceleration and hill climbing). The fuel cell's electrodes, donated by Union Carbide, had been used earlier in another ground-breaking fuel cell vehicle, General Motors' Electrovan, a six-passenger van that had been converted with Kordesch's help in 1967.

The Austin, which Kordesch acquired from a neighbor who had ruined the engine after driving it for 4 months without enough oil, was otherwise practically new. The fuel cell system, the batteries, the hydrogen tanks, and other auxiliary equipment made the car about 25 percent heavier, but its performance and range were adequate. The lightweight roof tanks, pressurized to 130–150 bars and holding up to 25 cubic meters of hydrogen, provided a range of about 190 miles. Refueling at a hydrogen-storage tank farm took only about 2 minutes. An electric motor with peak power of 20 kilowatts gave a top speed of about 50 miles per hour—fast enough for the placid suburb of Lakewood and its environs.

Normally, the CO_2 present in the atmosphere poisons the electrodes of alkaline fuel cells—one reason why transportation fuel cell developers have been concentrating on PEM-type fuel cells in the 1990s, since they are largely immune to the problem. Kordesch neatly solved that particular difficulty by installing a soda-lime air scrubber, which removed at least half of the air's CO_2 content of 0.03 percent. Additionally, his cell's alkaline (KOH) electrolyte absorbed some CO_2, reducing contamination even more. (Kordesch says the electrolyte could be easily changed when it had taken up too much CO_2.)

Kordesch could drive his Electro-Austin immediately after turning the key, using the batteries for power until the fuel cell reached its optimum operating temperature of 60–70°C within a few minutes. He increased the life expectancy of the fuel cell by shutting down the batteries between operating cycles, shutting off the hydrogen supply, and emptying the electrolyte into a reservoir; this exposed the hydrogen electrodes to air, effectively regenerating the catalyst and eliminating all parasitic currents.

Kordesch drove the little fuel cell car more than 13,000 miles in 4 years. In its practicality and usefulness, it was far ahead of its time. By some criteria, it still is. It was genuinely revolutionary, although few people would have noticed, or known, at the time. John Appleby, director of the Center for Electrochemical Systems and Hydrogen Research (CESHR) at Texas A&M University and one of the world's foremost authorities on fuel cells, devoted almost three pages to the Kordesch car in his seminal *Fuel Cell Handbook*.[1] "The significant feature of this vehicle was its practicality," said Appleby and his co-author, F. R. Foulkes. "It was essentially the project of an individual (though very knowledgeable) enthusiast, with access to the right materials. Kordesch produced a low-pollution vehicle with useful performance and range, with an easy-to-operate fail-safe system. The technology employed was quite old even in 1970, and present-day electrode construction would significantly improve vehicle performance and range."

Perhaps even more esoteric was an earlier Kordesch machine, a fuel-cell-powered moped fueled by hydrazine N_2H_4, a carbon-free nitrogen-hydrogen composite fuel best known as a propellant for small rockets used to position spacecraft or change their attitude with short bursts. Kordesch built it as a demonstrator for the US Army, which was looking at fuel cells for use in artillery radio service. A photograph reproduced in Kordesch's recent textbook[2] shows him in a helmet and a heavy parka, sitting on his moped on a midtown Manhattan sidewalk in front of the Union Carbide Corporation office building, with spectators kept at a safe distance by a velvet rope.

Union Carbide had always insisted that the moped and the car were Kordesch's private projects, with the company providing support only. The reason for this was a fear of liability. Kordesch said when he went to New York to demonstrate the moped on Manhattan's sidewalks he had to take out insurance totaling $2 million for 2 days to cover possible injuries and accidents. Fortunately, nothing ever happened.

The moped looked pretty normal except for a couple of metal boxes—the fuel cell and a NiCad battery—mounted in the frame ahead of and underneath the rider. Kordesch converted the Austrian-made Puch (bought at Sears, Roebuck) in 1966, and he drove it on public roads during his years at Union Carbide. There were two 16-volt, 400-watt hydrazine-air alkaline fuel cells and a Ni-Cad battery. The batteries could be switched in parallel or series for speed control. The range was about 60 miles on 2 liters of a 64 percent aqueous hydrazine, with an easy top speed of 25 miles per hour.

Kordesch was not the only person, nor was he the first, to try hydrazine for fuel cell power. Others had the same idea. In the 1960s, the now-defunct Allis-Chalmers Manufacturing Company built a small 3-kilowatt golf cart powered by that exotic fuel. Earlier, Monsanto Research Corporation developed a 20-kW hydrazine-air alkaline fuel cell system for a ¾-ton Army truck. In 1972, Shell Research in England put a 10-kW hydrazine-air system in a Dutch-built DAF-44 car. In 1982, according to Kordesch's book, the Japanese Shin-Kobe Electric Machinery Company employed the technology for a small 3-kW military power plant.[3]

Today it is difficult to pinpoint who should get the credit for the first practical fuel cell vehicle; however, a 20-horsepower tractor designed by W. Mitchell and demonstrated by Harry Ihrig of Allis-Chalmers in 1959 keeps cropping up in the literature as "one of the first." That machine was propelled by a 750-volt, 15-kilowatt, 917-kilogram hydrogen-oxygen alkaline fuel cell, a spinoff from space technology employing a circulating KOH electrolyte, according to Kordesch. The Allis-Chalmers tractor had just about all the characteristics of modern fuel cell technology: high-voltage bipolar stacks with porous metal electrodes, catalyzed with platinum and using an asbestos matrix to immobilize the liquid KOH electrolyte.

The following other fuel cell projects got underway in the early years:

• Union Carbide achieved a significant innovation in the mid 1960s with the development of all-carbon electrodes for the US Navy, and also for an experimental stationary 90-kilowatt hydrogen-oxygen prototype fuel cell (really a battery, according to Kordesch) for the Ford Motor Company.

• The availability of these new thin composite carbon-metal electrodes prompted General Motors to investigate fuel cells for transportation. That's how Kordesch got into the act. In 1967 GM decided to convert one of its six-passenger Handivans to a UC alkaline fuel cell system. The Electrovan's 32-module fuel cell system developed a peak power of 160 kilowatts—more than three times the output of demonstration PEM fuel cells that Ballard and International Fuel Cells were developing in the late 1990s for automotive use, but also a lot heavier. The Electrovan weighed 3400 kg, compared to about 1500 kg total for the standard version. According to Appleby and Foulkes, 1790 kg were directly due to the fuel cell and electric drive system, but somewhere between 450 and 680 kg were due to overdesign and test instrumentation. Still, the fuel cell van performed just about like its standard cousin, accelerating from 0 to 60 mph in 30 seconds and having a top speed of about 70 mph. The range was between 100 and 150 miles on liquid hydrogen, and 44–60 miles with compressed gases. It was a magnificent effort, despite a host of problems. These included excess weight and volume, short lifetime of costly key fuel cell components—the fuel cell lasted only about 1000 hours after activation—a lengthy and complicated startup procedure, system complexity, problems with temperature control, and, perhaps most important, safety: The potential for leaks of LH_2 and of the liquid KOH electrolyte, as well as the use of both liquid hydrogen and liquid oxygen in close proximity to each other in one vehicle are all just too dicey in case of a crash, dangers compounded by the system's high voltage (520 V). Still, GM concluded, according to Appleby and Foulkes, that "the rate of progress in this field and the strong advantages of fuel cells are sufficient incentives to maintain this effort."

• In Sweden, ASEA constructed a 200-kilowatt submarine unit in the mid to late 1980s; apparently some mishap prevented its completion.

• In Germany, the battery maker Varta and the electrical equipment builder Siemens assembled a demonstrator version of a fuel-cell-powered boat with an electric motor.

• In England, the Thornton-Shell combine operated a truck equipped with a hydrocarbon converter that extracted hydrogen from fossil fuel and cleaned it up with a shift reformer for use in an alkaline fuel cell.

A History of Fuel Cells

"At last, the fuel cell" was the headline of an upbeat three-page article in an October 1997 issue of *The Economist*. "A device that has been neglected for a century and a half is about to take its rightful place in industrial civilisation," said the subhead. The opening sentence read "Lovely technology, shame about the cost: That is the usual comment on fuel cells—a method of generating power that is 40 years older than the petrol engine." An accompanying editorial, "The third age of fuel," said "Just as coal gave way to oil, oil may now give way to hydrogen." In passing, the article mentioned that the principle of the fuel cell had been developed by an Englishman, William Grove, "a man who although he ended his career as a judge, began as a physicist."

Grove (1811–1896) was a professor of experimental philosophy at the now-defunct London Royal Institution and a friend of the famous physicist and chemist Michael Faraday (1791–1867), who discovered electromagnetic induction, invented the dynamo, and did research on electrolysis.

Grove, after experimenting with electrolysis, reasoned that it should be possible to reverse the process and generate electricity by reacting hydrogen with oxygen. In a classic experiment first reported in 1839, Grove built what is considered to be the first fuel cell. Though its electricity output was rather small, Grove was encouraged. Three years later he built a bank of 50 such cells employing dilute sulfuric acid as electrolyte. He also reported on a hydrogen-chlorine fuel cell, and he found that other carbon-bearing liquids—"other volatile bodies such as camphor, essential oils, ether and alcohol associated with oxygen"—"gave a continuous current."

Grove was quick to recognize the elegant symmetry of the processes of electrolysis and recombination of hydrogen and oxygen, but he soon abandoned work on a "gaseous voltaic battery" because his device was not able to produce enough power.

In the mid 1850s, attempts were made to develop a carbon-burning fuel cell. Nothing much happened until 1889, when, according to Appleby and Foulkes, a British scientist, Ludwig Mond, and his associate, Charles Langer, repeated Grove's earlier work. They were the first to call the device a fuel cell, and they tried to make it more practical by replacing oxygen with air and hydrogen with impure industrial gas obtained from coal by the so-called Mond-gas process. However, after achieving 1.5 watt at 50 percent efficiency, they dropped the project because of the high cost of the platinum catalyst, because the platinum was being poisoned by traces of carbon monoxide in the gas, and because of other problems. In 1896, an American engineer named J. J. Jacques constructed the largest system yet, which ran on coal and air at 82 percent efficiency, producing 1.5 kilowatts and (according to one account) operating for as long as 6 months at a time. Later, it was found that the efficiency was actually much lower because of a misreading of the chemical processes involved. Still, Appleby and Foulkes credit Jacques with being the first to think of a fuel cell as a device that could provide electricity for domestic use.

Francis Bacon's Contributions

The modern-day fuel cell technology that led directly to the machines used to provide electric power on the Space Shuttle began in 1932 with work done by Francis T. Bacon (1904–1992), an engineer associated with Cambridge University in England and a descendant of the renowned seventeenth-century philosopher-scientist Francis Bacon.

As a practical-minded engineer, Bacon reasoned that the high cost of the platinum catalysts employed in systems of the type developed by Mond

and Langer would never permit fuel cells to successfully enter the com
mercial marketplace. To get around this, Bacon decided to develop a
hydrogen-oxygen cell with an alkaline electrolyte and relatively inexpensive nickel electrodes. The temperature had to be raised to slightly more
than 200°C to make nickel sufficiently chemically active. This meant that
the cell had to be pressurized to keep the aqueous alkaline electrolyte from
boiling, which led to Bacon's discovery that pressurization made the cell
more efficient. (At first, Bacon pressurized his fuel cell to about 220 bars;
eventually, much lower pressures were used.)

Working first with an engineering company (C. A. Parsons and Co.) and
later at Cambridge University, Bacon built a single-cell unit in 1939. In
1946 he constructed an improved version of the single-cell design; it was followed by a six-cell device in 1954. In 1959, with help from the British government through the recently established National Research Development
Corporation, Bacon produced a 40-cell unit that delivered 6 kilowatts at
200°C and 38 bars pressure. In 1959, Bacon announced that he and his co-
workers had built and demonstrated a practical 5-kW unit with enough
power to run a 2-ton capacity forklift truck as well as a welding machine.
Two months later, Allis-Chalmers demonstrated the first fuel-cell-powered
vehicle, the famous 20-horsepower tractor mentioned earlier. In 1964, Allis-
Chalmers built a 750-watt fuel cell system for the Electric Boat Division of
General Dynamics to power a one-man underwater research vessel.

In the early 1960s, NASA discovered that fuel cells were suitable sources
of electric power for space flights of up to 14 days. Non-rechargeable batteries simply wouldn't last long enough, and there was no place in space to
plug in and recharge a NiCad battery. And fuel cells wouldn't need sunlight, as photovoltaic panels do. (In addition, photovoltaic panels require
backup batteries on low-orbit flights, since the spacecraft is sometimes in
the Earth's shadow.) Perhaps most important, a fuel cell system running on
liquid hydrogen and liquid oxygen produced about 8 times as much power
per weight—1.6 kilowatt-hours per kilogram, compared to 0.2 kWh/kg—
as the best batteries then available. NASA awarded more than 200 contracts for research on the physics, kinetics, electrochemistry, and catalysis
of fuel cell reactions, on how to manufacture electrodes, and on all sorts of
other details.

General Electric fuel cells with ion-exchange membranes were first used
on the Gemini orbital flights. The Apollo lunar missions used fuel cells

based on Bacon's design and developed by the Pratt & Whitney Division of United Aircraft Corporation (later United Technologies), which had bought licenses for Bacon's fuel cell technology. In 1970, United Technologies won the contract to provide fuel cells for the Space Shuttle program. In the 1990s, the International Fuel Cells Corporation, a subsidiary of United Technologies, was still supplying and servicing the fuel cells for that program.

NASA's massive engagement laid the groundwork for the renewal of interest in fuel cell development that came in the 1990s. "The massive US aerospace fuel cell effort has undoubtedly provided the single most important impetus to the development of electrochemical engineering science in respect to energy conversion," wrote Appleby and Foulkes. "There were many predictions that [fuel cells] would be the solution to the world's energy problems," they added. A host of programs got underway in the United States, in Europe (including what was then the Soviet Union), and in Japan to design and build various types of fuel cells—molten carbonate, solid oxide, phosphoric acid, alkaline, solid polymer, direct methanol— as primary electric power sources for utilities and for portable and small-scale uses.[4]

Utilities and Fuel Cell Power Plants

In the late 1960s, the Edison Electric Institute, United Technologies, and a group of electric utilities began to investigate the expected advantages of fuel cell power plants, with help from the gas utilities represented by the American Gas Association. Beginning in 1967, the AGA funded a 9-year program called TARGET (Team to Advance Research on Gas Energy Transformation), with United Technologies as prime contractor and the Institute of Gas Technology as subcontractor. The goal was to develop small natural gas fuel cells for home use, with phosphoric acid systems as the main technology and molten carbonate as a backup.

The hopes began to fade when certain technical difficulties became apparent to the fuel cell community in the late 1960s and the early 1970s. Coupled to a parallel slowdown in aerospace programs, these obstacles almost led to the demise of fuel cell development for terrestrial applications. Appleby and Foulkes list four major problems:

- Hydrogen was the only really useful non-exotic fuel, but using it with relatively inexpensive nickel catalysts in an alkaline fuel cell required high temperatures and pressures, costly pressure vessels, and ancillary equipment.

- Alkaline fuel cells required very pure hydrogen. That was problematic when hydrogen was produced from common fuels such as natural gas or coal. Any residual CO_2 in the hydrogen reacts with the liquid alkaline electrolyte, gumming up the electrodes' microscopic pores and slowing the overall chemical reactions.

- The use of "dirty" commercial fuels plus CO_2-containing air—as opposed to pure hydrogen and pure oxygen used on spacecraft—made the useful life of fuel cell systems (using construction materials commercially available at the time) too short for economical operation.

- In hindsight, it became clear that the close-knit community of fuel cell designers, engineers, and scientists had "tended to oversell the merits of the fuel cell before really having come to terms with all the teething troubles of an immature technology," Appleby and Foulkes wrote. "As a result of over-enthusiasm, deadlines were not met, and private funding for the fuel cell greatly declined. With the major budget reductions in the aerospace program during the early 1970s, government funding for fuel cells also slowed. As a consequence, fuel cell research and development almost came to a halt because the private sector was hesitant to assume the costs previously paid by the US government."

Interest in fuel cells on the part of many American electric utilities picked up again in the wake of the 1973–74 oil embargo. The government resumed funding development work for large-scale stationary fuel cell power plants. Overall, Appleby and Foulkes estimated, the government spent about $350 million (1986 dollars) between 1977 and 1984 on development of stationary fuel cells. A roughly equal amount was contributed by utilities and manufacturers.

Among the dozens of demonstration plants, subsystems, test facilities, and other related hardware constructed or planned in the United States, Europe, and Japan in those years, three examples stand out. The first was the first megawatt-class fuel cell power plant: the PC-19 phosphoric acid plant, built by United Technologies in South Windsor, Connecticut and tested during the first half of 1977.

In February 1977, as PC-19 began churning out electricity, the US Department of Energy, in cooperation with the Electric Power Research Institute, issued a request for proposals to build a big (4.5 megawatts) phosphoric acid demonstrator fuel cell power plant in New York City. Consolidated Edison was picked as the host utility.

The basic idea, Appleby and Foulkes say, was to demonstrate the feasibility of installing and operating a fuel cell power plant in an urban utility environment—as it turned out, a forbidding task at the time: "From the viewpoint of obtaining an operating license, the site had to be considered as one of the most difficult in the United States." In general, Appleby and Foulkes write, the history of the Manhattan fuel cell plant illustrates "the need for patience associated with the development of an emerging technology under real-world conditions." The plant was to be built within a year on a ¾-acre site at East 15th Street and Franklin D. Roosevelt Drive in lower Manhattan. It was to start operating in 1978, and the test was to be completed in 1979.

The plant was built, but it never produced any electricity. In addition to the usual squabbles among contractors and subcontractors and late deliveries, the builders and the utility had to contend with suspicion, fear, and ignorance of new technology on the part of New York's bureaucracy. For example, according to Appleby and Foulkes, the New York Fire Department, put in charge of overall public safety, decided initially that the plant was really a refinery and therefore would not be permitted in the city. The fire department later agreed to let it be built as an experimental power plant. A number of "stringent and unorthodox" safety tests resulted in irreparable damage to some one-of-a-kind advanced heat exchangers. To make matters worse, according to Karl Kordesch, water left in the heat exchangers during winter caused the exchangers to crack.

In early 1982 (3 years after the tests were supposed to have been finished) the fuel reformer ran into trouble because of clogging, which was attributed to the fire department's insistence on burning high-flashpoint naphtha fuel for safety reasons, whereas the plant had been designed for low-flashpoint fuel. In March 1983 the plant extracted the first hydrogen from naphtha for testing. In May the plant ran for 3 minutes on regular process gas on one occasion; a month later it ran for almost an hour. Verification of the transition from standby to load completed the crucial process and control test.

In the spring of 1984, all the fuel cell stacks were finally installed. It was then discovered that they didn't function properly. The stacks, which had been in storage for 7 years, were found to have had a limited shelf life because of the chemical interaction of the small amount of phosphoric acid electrolyte and the graphite bipolar plates between the cells. Design changes were required to prevent this from recurring. (Kordesch says they simply had dried out, something that could have been prevented by the addition of some extra phosphoric acid.) Ordering new stacks would have added another $11 million to the plant's cost, plus another year and a half, to a project whose construction cost (originally about $35 million) had already doubled. At that point, the partners threw in the towel. "The only logical step," according to Appleby and Foulkes, "was to terminate the project and clear the site, which was done by December 1985."

Although the Manhattan plant never produced any power, Appleby and Foulkes argue that the effort was not a waste and that it provided valuable lessons:

It passed all the other tests required to confirm the system viability. Most important, and in spite of the responsibilities and complexities involved in the codes at a major urban site, the plant showed that it was possible for a fuel-cell power generating station to comply with the demands of the local authorities.

After extensive extra testing under nondesign conditions, which resulted in considerable damage and long delays, it received its license to operate, and as a result, fuel cells will be the only new type of fuel-burning generator that will be allowed in Manhattan in the future.

In 1980, the Tokyo Electric Power Company (TEPCO), the largest privately owned electric utility in the world, signed a contract with United Technologies for a 4.5-megawatt plant of the same type as the New York plant to be built in Goi, in the Chiba Prefecture. This time the construction and licensing procedures were very simple, although special measures had to be taken to guard against possible earthquake activities, high tides, and oil spills. The TEPCO version incorporated a number of technical improvements, in part results of the New York experience. The first tests at half load (2 MW) started in April 1983, and the plant began operating at full power in February 1984, producing power at rates that exceeded design specifications. To be sure, there were some "teething problems," such as the fact that startup took 6–8 hours rather than the planned 4 hours because of some incorrectly dimensionioned plumbing. The plant operated successfully for 2 years on reformed methane, the longest run (500 hours) demonstrating dramatically improved reliability, according to Appleby and Foulkes. No deterioration in performance was seen in 2800 hours of operation.

The first Goi plant had been built within its budget of about $25 million—about 60 percent of the cost of the New York plant—and only 38 months after the order was signed. "Considering the problems that had occurred with similar subsystems in New York," Appleby and Foulkes wrote, "completion of the experiment in Goi must be regarded as a remarkable achievement, which leaves in no doubt the possibilities of properly designed fuel cells operating in a utility context."

Next, the International Fuel Cell Corporation began to develop an 11-MW plant for TEPCO, in cooperation with Toshiba Corporation, at Goi. Site preparation began in January 1989, and installation of the first of three groups of six stacks each, with 469 cells per stack, began in June 1990. The plant was started up on March 7, 1991, and it achieved an output of 11 MW output before the end of April. By August, the plant had produced 10,264 MWh of electric power in 1414 hours of operation, including one uninterrupted stretch of 875 hours. Initial evaluations found that the measured efficiency exceeded the design value and that the stacks were in "extremely good condition," according to an interim report.[5] Some of the fuel cells leaked as a result of an electrochemical reaction between the reformer exhaust gas and the carbon electrode material, some of the leaks being "dozens of times greater" than the values recorded in acceptance inspections, but others showed almost no leakage. Several other breakdowns occurred in this phase (as is normal when a new technology is being developed), none of them serious. The plant, intended only as a technology demonstrator, operated until March 1997, producing more than 77,000 MWh in more than 23,000 hours of operation.[6]

Around 1993, United Technologies, through its International Fuel Cells Corporation and ONSI subsidiaries, began building 200-kW phosphoric acid fuel cell plants. About 200 of these plants have been deployed, around the world, at a rate of about two dozen a year since the program started. So far, these are the only commercially available fuel cell plants.

As Appleby and Foulkes observe, it was once widely assumed, on the basis of the early space-related efforts, that small fuel cells (putting out a few kilowatts) for electric vehicles would come first, and that larger units for stationary power generation would follow; however, in the late 1980s it seemed that large, multi-megawatt stationary systems would come into commercial use first. "Only a large assured market can justify the major public and private expenditures necessary to ensure a proper developmental effort," Appleby and Foulkes argue. Another reason, they say, may be that big stationary fuel cells do not face the resistance of strong lobbying groups (such as automobile manufacturers, who "might prefer to use the internal-combustion engine for as long as is profitable").

Two molten carbonate fuel cell demonstration plants have been started up in California (one in Santa Clara and one at the Miramar Naval Station, near San Diego). And developers of other technologies (e.g., solid oxide fuel cell power plants) are confidently talking about commercialization within a few years after 2000. In mid 1999, Siemens-Westinghouse was busy gearing up to build four 250-kW solid oxide fuel cell demonstration plants, two in Europe and two in the United States.[7]

Efforts to commercialize home-size fuel cells—in part an outgrowth of the deregulation drive in the electric utility industry in the mid 1990s—resumed in the second half of the 1990s. These units were to be capable of operating on readily available natural gas and also on propane or other fuels, which could be delivered to islands and other remote areas. A California-based group organized by US and Canadian utilities, the Small Scale Fuel Cell Group, was apparently the first to issue a Market Opportunity Notice for such systems in 1997, essentially asking for bids and guaranteeing purchase of a dozen or so units for demonstration purposes and eventual mass production. The Boston-based American Power Corporation[8] had teamed up with the Electric Power Research Institute to pursue similar goals. Other developers jumping into the competition included Plug Power (a New York State company founded in the summer of 1997 with help from a Detroit Edison affiliate), H Power, and De Nora of Milan. The Swiss company Sulzer and the Canadian firm Global Thermoelectric announced development plans for small solid oxide fuel cell systems.

Even further down the scale are tiny fuel cells that their developers say will power cellular telephones, laptop computers, and electric hand tools. A fuel cell developed by a former Los Alamos weapons specialist, Robert Hockaday, was claimed to be capable of giving about 50 times more talk time when used in a cellular phone than conventional Ni-Cad batteries at what Hockaday hoped would be half the cost. Manufactured using lithographic etching techniques of the kind used in the manufacture of computer chips, it would be refueled with a squirt—about an ounce and a half—of methanol. The device, announced near the end of 1997, was initially designed to produce only enough power (about 4 watts) for a cell phone, but Hockaday expected to scale it up to produce the roughly 30 watts required by a laptop computer.

Another power source for cell phones was being developed by a scientist at Motorola, Christopher Dyer. Dyer's thin-film fuel cell would generate power from a gas mixture containing both hydrogen and oxygen (whereas the usual systems need separate gas supplies)—something which Dyer said in a 1999 article[9] should result in a simpler and cheaper system. Dyer, who first reported on his work in 1990 when he was still at Bell Communications Research,[10] described an extremely thin (less than a millionth of a meter) gas-permeable electrolyte sandwiched between two thin layers of platinum.

In 1996, the DAIS Corporation of Palm Harbor, Florida, reported that it had developed a new, low-cost membrane material and had built small PEM fuel cells (in the 40–125-watt range). Larger devices that would produce up to a kilowatt were said to be targeted for future development. A small 5-watt prototype cell shown at conferences in 1996 was a little bigger than a golf ball. The membrane material was developed initially at the Rensselaer Polytechnic Institute in a program headed by Gary Wnek, chairman of RPI's chemistry department and director of its polymer science and engineering program. In 1999, DAIS merged with Analytic Power to form the DAIS-Analytic Corporation

Fuel Cells: How They Work

Fuel cells are directly related to conventional "primary" batteries (in which a zinc casing reacts with manganese dioxide to produce electricity) and to rechargeable "secondary" batteries (which use lead and lead dioxide or nickel and cadmium to store electricity). But whereas a primary battery simply runs down after its zinc casing is exhausted, and a secondary battery must be periodically recharged, a fuel cell keeps running as long as it is supplied with fuel (e.g., hydrogen), just as an internal-combustion engine or a gas turbine will run as long as it is fueled.

A single fuel cell is essentially an "electrochemical sandwich," only a fraction of an inch thick, with a negatively charged anode on one side, a positively charged cathode on the other, and an electrolyte (a watery acidic or alkaline solution or a solid plastic membrane that permits the migration of electrically charged hydrogen atoms from the anode to the cathode) in the middle. Many individual cells can be "stacked" to produce a usable amount of electricity.

Fuel cells operate almost silently,[11] generating low-voltage direct current from the catalytically aided electrochemical reaction between a fuel (hydrogen or a hydrogen-rich carrier such as natural gas) and an oxidizer (oxygen, either taken from the ambient air or carried in separate storage vessels). Whatever sound is associated with them comes from auxiliary equipment: pumps to transport the fuel through the system and to remove the water produced in the reaction, a blower or compressor to cool the stack and to convey oxygen to the fuel cell, and perhaps a humidifier (required by some types, such as PEM cells).

Practical fuel cell systems are very complex to design and build—especially small, rugged ones for cars, trucks, and buses, which must stand up to bumps and to temperature variations. (This is one reason why it took so long to put the first prototypes on the road.) But the basic principle of how a fuel cell works is fairly straightforward.

In a hydrogen-oxygen fuel cell with an acid electrolyte, molecular gaseous hydrogen is fed in on the anode side. There electrons are stripped out of the H_2 molecules, leaving positively charged hydrogen ions. These migrate through the electrolyte layer to the positively charged cathode, where they combine with oxygen to form water. Meanwhile, the liberated electrons (electric current) flow from the anode through a wire or some other metallic material back to the cathode, performing work—for example, operating an electric motor, lighting up a light bulb, or powering a cell phone.

Materials other than hydrogen can serve as ionic energy carriers too—for example, carbon (as a carbonate ion) and oxides of various elements. "In theory, any substance capable of chemical oxidation that can be supplied continuously (as a fluid) can be burned as the fuel at the anode of a fuel cell," Appleby and Foulkes explain. "This overall reaction may be viewed as the cold combustion of hydrogen with oxygen"—cold combustion because it takes place at temperatures much lower than those in a conventional open-flame process. In normal combustion all the energy generated is released as heat. In a fuel cell, however, part of the free energy of this electrochemical reaction is released directly as electricity; only the remainder is released as heat—obviously a much more efficient process for powering electric equipment and electric cars.

An important aspect, one that has assumed much greater importance in recent decades because of the concerns over global warming, is that the preponderant nitrogen component in our atmosphere—about 80 percent of the air we breathe is nitrogen—plays no part in cold combustion. Only hydrogen and oxygen react with each other in a fuel cell. By definition, this process precludes the production of nitrogen oxides, a key component in air pollution. In contrast, any open-flame process (such as the combustion of gasoline in conventional car, bus, and truck engines and jet turbines) produces these harmful emissions.

There are several basic types of fuel cells, differentiated by their electrolytes and by the temperature ranges in which they operate[12]:

alkaline   This is the type favored by Kordesch and other early pioneers. Alkaline fuel cells are still used on the Space Shuttle today (in highly refined designs that have evolved to levels of unparalleled efficiency and low weight). They operate at 60–90°C. They have efficiencies of 50–60 percent in converting chemical fuel to electricity. The electrolyte is 35–50 percent KOH. The principal applications are in space, but at least one company (ZEVCO, renamed ZeTek in 1999) thinks they are still the wave of the future for road transport, because they are cheaper and because they don't require platinum as a catalyst. AFCs' main disadvantage is that ambient CO_2 can degrade their performance. However, as Kordesch demonstrated with his Austin, there are simple, economical industrial technologies for purging CO_2 from alkaline systems.

proton exchange membrane[13]   PEM fuel cells operate between 50°C and 80°C. Their efficiencies range from 50 percent to 60 percent. They use polymer-type membranes of the Nafion type (produced for more than 20 years by DuPont, more recently also produced by Ballard, W. L. Gore, Asahi Chemicals, and others). More advanced high-temperature membranes that would deliver higher performance are under development. Major international carmakers are pinning their hopes—and spending buckets of money—on these in the race to develop a viable fuel cell engine for cars and buses. PEM-type systems are the fuel cells of choice for transportation because of their fast startup (they can produce usable amounts of power almost instantly, even when close to the freezing point of water), their high power density, and their relative ruggedness.

phosphoric acid   Typically designed for stationary power applications, phosphoric acid fuel cells operate with about 55 percent efficiency at temperatures of 160–220°C. The electrolyte consists of concentrated phosphoric acid. Until the late 1990s the PAFC was the only type that was commercially viable. For a few years, the sale of International Fuel Cell PAFCs has been supported by the US government with a small subsidy program (some have described it as "pork barrel") as a sort of icebreaker to induce utilities and other commercial users to accelerate the phasing in of fuel cells of all types in the years ahead. The fuel of choice is natural gas.

molten carbonate   Molten carbonate fuel cells, which run at 620–660°C, are considered promising for future power-generation needs. Their efficiency is typically in the 60–65 percent range. As the name implies, the electrolyte is a melt (a binary alkaline carbonate mixture). Test operations by the Santa Ana, California utility of a 2-megawatt MCFC power plant built by Energy Research Corporation (renamed FuelCell Energy in 1999), and of a plant built by MC Power at the Miramar Naval Station, were in the forefront during the latter half of the 1990s. Those plants were fueled by natural gas, but propane and other fuels may also be useful.

solid oxide   Solid oxide fuel cells are expected to come into their own as power generators for utilities shortly before 2010. They operate at the highest temperatures yet: 800–1000°C. Their efficiency is in the 55–65 percent range. The electrolyte typically consists of an alloy of rare earth materials, such as yttrium-stabilized zircon dioxide, a ceramic material. Small SOFCs are under development for a variety of applications, for home use, for various uses in remote locations (e.g., to power pumps for long-distance natural gas pipelines), but also as onboard power generators for automobiles (taking the place of traditional generators and alternators). Like the molten carbonate type, they also run on natural gas.

Further down the road are more advanced designs, the two most notable being the Direct Methanol Fuel Cell (DMFC) and the Reversible or Regenerative Fuel Cell.

Probably the best-known developers of DMFCs in the international fuel cell community are teams of scientists at the Jet Propulsion Laboratory, at Ballard Power Systems, at the Los Alamos National Laboratory, at Siemens, and at the University of Newcastle. The basic idea is to develop a type of fuel cell—an offshoot of PEM technology—that could be fed directly with methanol or a solution of methanol in water. This would be an advance over current types, in which hydrogen must be extracted from methanol in a miniature chemical plant (a "steam reformer" or a partial oxidation processor) before it can be fed to a fuel cell. That approach had technical drawbacks, such as the slight contamination of the produced hydrogen ("reformate" in engineering parlance) with traces of hydrocarbons. Developing a fuel cell robust enough to take methanol straight would "significantly lower system size, weight, complexity, and temperature than in existing fuel cell systems," said Gerald Halpert, lead scientist of the JPL effort, in an autumn 1996 paper delivered at a conference in Germany. At a February 1999 methanol conference in Washington, an ebullient Halpert said "the technology is ready for commercialization, ready to go." He said JPL's device had achieved 34 percent stack efficiency at 90°C, that it had operated continuously for 200 hours with no noticeable degradation, and that it had operated for 4500 hours intermittently at 60°C with minimal loss of power. Others were not so sanguine. Speaking at the same conference, Ferdinand Panik, head of DaimlerChrysler's fuel cell project, said that DaimlerChrysler was working "very intensively" on direct-methanol fuel cells but did not believe that they were nearly as efficient as reformer-equipped systems. Nor did Panik think that direct-methanol fuel cells would catch up anytime soon. "My feeling," he said, "is [that the DMFC] is 8–12 years behind" processor systems.

Reversible fuel cells—systems that can be switched between producing electricity from the cold combustion of hydrogen and oxygen and splitting water again into hydrogen and oxygen by electrolysis—were first developed some years ago as laboratory exercises by the Hamilton Standard Division of the United Technologies Corporation.

Proton Energy Systems (four of whose founders came from Hamilton Standard and worked on this technology for most of their careers there) has begun development of a reversible fuel cell. In the summer of 1999, Proton Energy Systems announced that it had successfully tested a "Unitized Regenerative Fuel Cell"—in effect, the company said in an announcement, "a high-performance large-scale battery without many of the limitations of a conventional battery." The concept was born in the early 1990s as a part of a NASA plan for a huge, unmanned, solar-powered flying wing. In order to stay aloft for months, the flying wing would have to store and recycle solar energy for night flight. The goal was to have the craft stay aloft for months, at stratospheric altitudes of 70,000 feet or more, for military and civilian purposes such as long-term weather observation. The energy needed to crawl up to the fringes of space (a 1993 story said that the plane would climb at a rate of only about 12 knots, requiring about 6 hours to reach 70,000 feet, and would then cruise at only 27–28 knots) would come from solar cells covering about 75 feet of the 100-foot wings' upper surface. Operating as electrolyzers during the day, these onboard fuel cells would split water and would then use the hydrogen and the oxygen to produce electricity to turn the propellers at night. The entire system has not flown so far: the original "Pathfinder," which reached an altitude of more than 71,000 feet in the summer of 1997 during a 14-hour flight, relied on battery power for energy needs such as avionics and landing lights. The electricity to churn the plane's six propellers came only from solar cells. Later generations, such as the even bigger, planned "Helios" 200-footer were expected to incorporate these reversible fuel cells and storage tanks to achieve the goal of staying aloft for months at a time.

Almost from the beginning, the designers of the reversible fuel cell felt that it could have automotive applications. James McElroy, program manager for electrochemical systems at Hamilton Standard, said in a 1993 interview that a reversible fuel cell "would be like any other chemical fuel cell, except that you wouldn't need a new infrastructure (for fuel). You would plug it into the electric power grid as infrastructure." In a car, it could also be used for regenerative braking and for regenerating hydrogen, he added. However, McElroy cautioned, use in ground transportation was much further off than use in some esoteric, high-flying solar airplane. "It's not nearly as mature," he said at the time. For one thing, the envisioned NASA system would be simpler and more efficient, since it would use both pure hydrogen and oxygen generated onboard by electrolysis. An automotive system would presumably use air, which, with its oxygen content of only 20 percent, would be less efficient. Also, the use of compromise catalysts would, presumably, make it less efficient than one-way electrolyzers or fuel cells.

Is there likely to be one dominant fuel cell technology? It is hard to say, but it seems doubtful that any particular system will emerge as the sole winner for all applications. Most likely, different types of fuel cells will be built for different purposes, with bottom-line economics as the determining factor. As the twentieth century wound down, PEM fuel cells were getting most of the attention in the accelerating international race to develop a new power plant for cars and buses—and they were getting most of the research funding, both governmental and corporate. But a minority, including Karl Kordesch, Nick Abson (founder and chief executive of ZEVCO and ZeTek), and some others, remain unconvinced. They still believe that alkaline fuel cells will ultimately emerge as the superior choice. Kordesch, who near the end of 1999 continued to make "good progress" with alkaline fuel cells (including one version that uses ammonia as fuel) and with direct methanol fuel cells, says that much of the attention PEM technology has been getting is "hype." Alkaline fuel cell technology, he said in an e-mail, "is not modern enough for the PEM people"; however, it is "affordable," and it can be "completely shut down without [fear] for the safety of the garage and surroundings"—the latter referring to lingering concerns over venting of hydrogen from gaseous-hydrogen or liquid-hydrogen tanks in enclosed spaces and the attendant conflagration risk. "Economy," said Kordesch, "will decide."

8

Hydrogen in Aerospace: Clean Contrails and the Orient Express

On April 15, 1988, an event occurred that, though largely ignored by the American media, was a landmark in the history of aviation and of hydrogen energy: A modified Tupolev 154 airliner,[1] re-named TU-155, took off from an airport near Moscow on a test flight with the turbofan engine on the right side operating on liquid hydrogen. The English-language service of the Soviet news agency TASS reported that the engine could be operated either on liquid hydrogen or on liquefied natural gas. Hydrogen, the story hinted, was expected to be the long-term fuel of choice for global aviation. That evening, the Soviet television news show *Vremya* showed the plane taking off and landing. In the United States, a brief segment was distributed by the CBS television network to local affiliates, but apparently few stations bothered to show it. The cable channel CNN showed footage of the flight 3 days later. The Washington publication *Defense Daily*, which did carry a brief story, reported that the flight had lasted 21 minutes. The *New York Times* noted the event about 4 weeks later.

To interested observers, it appeared that the Soviet Union was gearing up to develop hydrogen-fueled aviation and that it had a leg up over the West in exploring this new technology. Maybe it was only old-style communist propaganda, but both TASS and the party newspaper *Izvestia* implied that aircraft powered by liquid hydrogen and liquid natural gas were going to be the wave of the socialist future.

Even more startling to the international hydrogen community was that this was not some sudden public relations ploy. The Soviets had been at it for a long time. Alexei Tupolev, the son of the world-famous aviation engineer who had created the series of Tupolev airliners and a brilliant aircraft designer in his own right, was quoted by *Izvestia* as saying that preparations for the flight had been going on for 9 years and had been a priority for Soviet aviation research.

G. Daniel Brewer, a now-retired aeronautical engineer who had headed Lockheed's hydrogen program in the period 1972–1984, commented that the Tupolev team's achievements placed it ahead of US efforts to develop the "National Aerospace Plane" (a hypersonic hydrogen-fueled experimental craft, announced by President Reagan 2 years earlier, that was to fly from a normal runway to beyond the atmosphere and back). Brewer told the *Defense Daily* reporter that the Russians might be "five years ahead of us in a technology that we need to develop." "Since the Russians have flown [their] plane," he said in another interview, "it possibly means that they have done all the [cryogenic] valves, heat exchangers, pumps. . . ." On the other hand, "if the plane used a pressure-fed liquid hydrogen system as we did in the late 1950s, it wouldn't be all that great," he said. (He was referring to the fact that in 1957 an American B-57 twin-jet bomber had flown many experimental missions, of about 20 minutes each, with one engine operating on liquid hydrogen.) But "if they have that system to take off, fly and land (on pump-fed liquid hydrogen) that would really be a step forward," Brewer added. "It would take us at least five years to develop it and put it in operation to achieve the same point."[2]

At the World Hydrogen Energy Conference held in September 1988 in Moscow, Tupolev told an overflow audience that hydrogen had been used exclusively to power the right engine of the TU-155 throughout the entire flight cycle—takeoff, cruise, and landing. Tupolev said that the liquid hydrogen had been carried in a removable stainless steel tank that had taken up most of the plane's rear compartment, and that there had been enough for a flight of up to 90 minutes. Potential ignition sources—electrical wires, hydraulic lines, conventional fuel lines—had been removed and relocated elsewhere in the fuselage. Pressurized air had been blown into those areas and into the space between two walls that separated the rear compartment from the rest of the plane to prevent the buildup of potentially dangerous air-hydrogen mixtures.

The Soviet feat caused muted anguish in Washington among the few lawmakers who supported the development of hydrogen energy technology. Tupolev's triumph was "yet another case in which the US had stood idly by in a critical area of research and allowed other nations to move far ahead,"

Representative George Brown Jr. (D-California), a member of the House Science, Space and Technology Committee, said shortly after the flight. "The Soviets are not alone in the development of liquid hydrogen as a fuel for their transportation. Now, with the Soviet cryogenic program, we face a real national security threat. Judging by the lack of support for hydrogen R&D in this country, it appears that the United States will be forfeiting to other more farsighted nations."

Two weeks later, Senator Spark Matsunaga, Brown's counterpart in the Senate and a frequent co-sponsor of hydrogen legislation, called the Soviet flight "a milestone." In an April 27 speech on the Senate floor, Matsunaga urged Congress and the Reagan administration, both monumentally uninterested in cleaning up the environment or in alternative renewable energy, to support hydrogen legislation that he and Brown had introduced earlier and which had been languishing in committee. "It is not too late to move on this bill and pass it before the end of this Congressional session," Matsunaga said. "As the news from Moscow indicates, my bill will come none too soon for international competition in aviation and space."

Thirty years earlier, Sputnik had "shocked and galvanized our nation," Matsunaga told his colleagues. "Eleven days ago, the Soviets announced another scientific 'first' which, curiously enough, has gone virtually unnoticed to date in this country," he said. The flight of a jet plane "powered by liquid hydrogen in an engine modified to accept cryogenic fuel . . . spells the advent of cryogenic aviation and represents a milestone in the march toward a hydrogen economy." Given the importance that the current administration has placed on the development of a "trans-atmospheric aircraft" and "the recognition that hydrogen would be the fuel of choice for such a craft," Matsunaga said pointedly, "it is a mystery to me why the administration has neither welcomed nor supported my legislation. Perhaps now that the Soviets appear to have again stolen a technological march on us in this regard, administration officials might be moved to reconsider their position on my bill."

Presumably unbeknownst to Senator Matsunaga when he made his speech, an American aviation pioneer and veteran pilot was quietly getting ready to claim the hydrogen aviation spotlight (or at least a small part of it) for the United States. Working almost alone, William Conrad, an octogenarian retired FAA-certified air transport ratings examiner who had started flying in 1929 and who had been Pan American's first director of flight training, was at work in a hangar at Fort Lauderdale's Executive Airport, preparing a second-hand four-seat Grumman Cheetah to take off, fly, and land while running exclusively on hydrogen.

On Sunday, June 19, 1988, Conrad, a recipient of the Wright Brothers Memorial Award and an inductee in the Aviation Pioneers Hall of Fame before this flight, got up at 5 A.M., drove some 60 miles north to West Palm Beach, filled his custom-made 40-gallon portable cryogenic tank with liquid hydrogen purchased at $2.80 a gallon from the commercial supplier Tri-Gas, and drove back to Executive Airport. There, he and his mechanic used a mobile crane to lower the tank into the cockpit behind the pilot's seat and connected the fuel lines to the plane's custom-made hydrogen fuel system. Inspectors from the National Aviation Association and the International Aviation Federation were on hand to witness the event; they certified that the gas tanks in the wings had been sealed off. Conrad revved the engine and checked it out "to make sure that everything was all right," strapped himself in his seat, slid the cockpit canopy shut, and was ready to go. He then received clearance from the tower for what was supposed to be only a high-speed taxi run. (As he related later, fear and ignorance of hydrogen was such that he never managed to get permission for a full test flight.) At about 1:30 P.M., Conrad pushed the throttle lever full forward. The Lycoming E2G engine, rated at 150 horsepower on gasoline, produced about 10 percent more power on hydrogen.[3] About 20 seconds later and some 600–700 feet down runway 8, the Cheetah became airborne. It rose about 300 feet, then touched down safely. The flight had lasted 39 seconds, never leaving the airspace above the 8000-foot runway. "The damn thing just got into the air with me," the tall, lanky Conrad deadpanned afterward. Even though the hop was technically only a high-speed taxi test, it was enough for NAA and FIA inspectors to certify a "world first" for Conrad.

Conrad, who had spent about $100,000 on the project (including $26,000 of his own money for the plane itself), said that he "just wanted to establish the fact that the United States has the first plane to run solely on hydrogen." The purpose, he added, was "to get a little publicity in order to make people realize that hydrogen is available and that it is a non-polluting form of energy."[4]

President Reagan sent Conrad a congratulatory letter later that year. "Your remarkable flight last June was a milestone in aviation history, and your 14 years of preparation for it were welcome proof of the vitality of our search for new energy sources to improve transportation efficiency and benefit the environment," Reagan wrote.

Conrad died just about a year after the flight, on June 27, 1989, after a long bout with cancer. His passing was noted by Senator Matsunaga "with a heavy heart" in a September 19 speech in which he urged his fellow senators to support research on the use of hydrogen as an aviation fuel. "We must not lose the initiative in hydrogen-powered flight that Bill Conrad set for our country," said Matsunaga.

In 1957, an American B-57 twin-engine jet bomber cruised at about 50,000 feet over Lake Erie at about Mach 0.75. It looked just like any other B-57, except perhaps for the wingtip tanks: the one on the left was slightly thinner and less smoothly rounded than the one on the right, and it looked a bit makeshift. Another difference, almost unnoticeable, was a small boxlike structure mounted outboard of the left engine underneath the wing. It was an air-hydrogen heat exchanger.

Based at the NACA Lewis Research Center, near Cleveland, this B-57 was the first airplane ever to fly partially powered by liquid hydrogen. After taking off on kerosene, the pilot could switch one engine to draw pressurized liquid hydrogen from the left wingtip tank. Pressure provided by helium carried in the right wingtip tank would push the liquid hydrogen to the heat exchanger, where heat from the inrushing air would turn it into a gas. The hydrogen gas could then be burned quite normally in the left engine.

The plane flew for about two years, using hydrogen in flights as long as 17 minutes, always at high altitudes and at cruising speed, with no fuss and no problems.

Also in the 1950s, studies conducted by the Lockheed Corporation in collaboration with the engine maker Pratt & Whitney and a division of AiResearch led to a remarkable episode in the annals of hydrogen technology: the secret project to develop a liquid-hydrogen-fueled supersonic spy plane, the CL-400 (code-named Suntan). The story of this project was closely guarded until 1973, when former project engineer Ben Rich described the never-built plane (it never reached the prototype stage) to an audience of aviation and hydrogen experts at a hydrogen aircraft symposium at NASA's Langley Research Center. That was the first time anything about this project, which eventually led to the famed SR-71 "Blackbird," was described to outsiders.

As Rich told it, the CL-400 was intended to fly at close to 100,000 feet, to cruise at Mach 2.5, and to have a range of 1100 miles. The first of two prototypes was to be delivered only a year and a half after the go-ahead. The CL-400 was supposed to carry a crew of two and a payload (mostly aerial cameras and other reconnaissance equipment) of only 1500 pounds. More than 164 feet in length, it was to have a wingspan of about 84 feet. The fuselage was to be 10 feet in diameter. The two engines, each delivering 9500 pounds of thrust, were to be mounted at the wingtips, where some other planes carried auxiliary fuel tanks. Small outrigger landing gears at the wingtips would stabilize the plane on the ground. "The low density of the cryogenic fuel necessitated a large fuel volume," Rich recalled—21,404 pounds of liquid hydrogen. At takeoff the plane would have weighed 69,955 pounds.

"There was very little knowledge of liquid hydrogen or cryogenic handling, other than that related to the hydrogen bomb experience," Rich said at the Langley symposium. "Consequently, it was necessary for us . . . to set up a test facility to learn how to handle liquid hydrogen in a fashion no different from hydrocarbon fuel. It was our feeling that if you could not handle liquid hydrogen like gasoline, you did not have a practical vehicle." Under the overall direction of Clarence ("Kelly") Johnson, the director of Lockheed's advanced projects organization, the "Skunk Works," Rich and his crew set up small-scale hydrogen tanks and fuel supply systems to get the hang of handling the stuff. Much effort was spent on discovering how to run liquid hydrogen at −425°F through wings heated to several hundred degrees by air friction at supersonic speeds.

A great deal of work also went into studying the much-feared but largely unknown hazards of hydrogen, including a large number of deliberate attempts to explode it. "The deflagrations were generally mild due to the high hydrogen flame velocity," Rich said. "The fireball was much less than a comparable kerosene fire. . . . Only 2 of 61 liquid hydrogen tests produced bona fide explosions. In both cases oxygen was deliberately mixed with liquid hydrogen"—just about the most dangerous situation imaginable and not very likely in normal operations. Said Rich: "We showed that a hydrogen aircraft was feasible. Liquid hydrogen could be handled, with the proper procedures and care, as easily as hydrocarbon fuel."

The CL-400 got as far as wind-tunnel testing, component development, and procurement of basic materials, but in 1957 the program was canceled. A technical reason for the cancellation was that, once the overall design was fixed, there was not much of an opportunity to improve the plane's range, mainly because of the peculiar characteristics of liquid hydrogen. "The airplane was too short-legged and had not more than 5 percent potential range performance stretch," according to Rich—the result of the design's low supersonic lift-to-drag ratio. "It grew into a big dog without enough range," Rich noted in a 1994 *Popular Science* interview. Another reason for the cancellation had to do with the logistics of transporting liquid hydrogen to air bases on the periphery of the Soviet Union and China. "How do you justify hauling enough LH_2 around the world to exploit a short-range airplane?" Rich asked. Other studies showed that switching to hydrocarbon fuels could double the plane's range; ultimately, the decision to do so led to the SR-71.

According to John Pike, the aviation expert of the Federation of American Scientists, although "Suntan" was a failure, "the work done on its hydrogen propulsion system laid the groundwork for subsequent application of this technology for space rocket propulsion." The "Suntan" program led to the Centaur (the first space rocket fueled with liquid hydrogen), the Apollo lunar program, and the Space Shuttle. Eventually it may lead to a new generation of aerospace planes.

Pound for pound, liquid hydrogen stores about 2.8 times as much energy as jet-grade kerosene. At the same time, however, a pound of liquid hydrogen takes up three to four times as much volume as a poundof jet-grade kerosene. A pound of LH_2 has a heating value of 51,500 Btu. Jet-A, the most common jet fuel, has only 18,600 Btu per pound. But a cubic foot of liquid hydrogen has only 227,700 Btu, whereas the same volume of Jet-A has 906,000 Btu. Thus, to obtain the same heating value contained in a cubic foot of Jet-A requires 3.97 cubic feet of liquid hydrogen. For this reason, hydrogen-fueled aircraft will have to have "fat" fuselages; however, they will be lighter than kerosene-fueled planes.

Quite apart from the environmental advantages of burning hydrogen in jet engines (which are basically the same as those of burning hydrogen in car engines: no smoke, no unburned hydrocarbons, no CO_2 or CO, low NOx emissions), there may be economic advantages.

At the same 1973 NASA hydrogen symposium at which Ben Rich spilled the beans on the "Suntan" spy plane, two NASA scientists, Cornelius Driver and Tom Bonner Jr., presented first rough estimates of what hydrogen use might mean in terms of a practical design goal: a jet freighter capable of carrying 265,000 pounds over 5000 nautical miles. Their study did not assume any design advantages that might come about as the result of using hydrogen; they merely substituted liquid hydrogen numbers for kerosene data in standard design computations. Nonetheless, the numbers were startling. Using the general formula that provides a measure of an airplane's overall flight efficiency,

$$\frac{M(L/D)}{\text{SFC}},$$

where M is Mach number, L is lift, D is drag, and SFC is specific fuel consumption, Driver and Bonner said that using hydrogen in planes such as the 707, the DC-8, the 747, the DC-10, and the L-1011 would roughly double fuel efficiency "if no serious problem is encountered in providing the necessary hydrogen volume."

In 1976, G. Daniel Brewer came up with more refined data on both supersonic and subsonic hydrogen planes. Of 24 designs that had been studied, Brewer discussed a long-range subsonic passenger plane (400 passengers, range of 5500 nautical miles) as a representative example.[5] The data were very encouraging. Gross weight was 177 tons for the hydrogen plane, versus 237 tons for the Jet-A version. Fuel weight for the H_2 plane was one-third lower, and the engines could be smaller and lighter since they would have to deliver 11 percent less thrust. Wing area was reduced by about one-fourth because of the lighter loads, though the hydrogen plane was about 10 percent longer. The plane would carry liquid-hydrogen tanks inside the fuselage fore and aft,[6] with no direct access from the passenger compartment to the cockpit. Because the forward hydrogen tank would be in the way, a potential hijacker would be unable to reach the cockpit in flight. The hydrogen-fueled subsonic transport would be considerably quieter during takeoff and cruising but not during landing, the analysis predicted. For fly-over noise, the hydrogen plane would register 104.9 EPNdB (effective per-ceived noise level in decibels); the Jet-A plane would register 107 EPNdB. On a landing approach, the LH_2 plane would be somewhat noisier: with most of its fuel gone, a hydrogen plane would weigh about the same as the similarly empty Jet-A plane; however, because it would have smaller engines and smaller wings, the engines would have to run at a higher power setting to maintain the same glide angle, which would make more noise. The LH_2 plane would affect a smaller area with high noise—the so-called 90-EPNdB contour, the area around the runway in which these noise levels are regis-tered during takeoff and landing—than an equivalent kerosene-fueled plane. The hydrogen version would produce no carbon oxides, unburned hydrocarbons, or smoke, but it would emit water vapor and some nitrogen oxides. In cruise conditions, Brewer estimated, hydrogen engines would produce about twice the amount of water vapor than the kerosene plane—82.4 pounds per nautical mile, versus 41.9 pounds. Would these water-vapor emissions affect the weather? In the absence of real experimental data it was hard to say, but Brewer felt that the meteorological impact would be minimal or nonexistent. If the water vapor coming out of the plane's four jet engines were visualized as a thin film of water the width of the engines' exhaust nozzles, the thickness of this water film would be only 0.00008 inch, and 82.4 pounds of water spread over a nautical mile doesn't seem very worrisome.

In 1989, C.-D. Schönwiese of Frankfurt University's Meteorological Institute postulated that high-flying hydrogen-powered aircraft may indeed add to the greenhouse effect, but hydrogen was unlikely to cause problems with planes flying in the lower atmosphere. Schönwiese, who spoke at a 1989 meeting on hydrogen energy technology organized by the Verein Deutscher Ingenieure, said that in theory the generation of steam, a very "climate-effective" trace gas, could be problematical. However, he explained, steam exists normally near the ground, concentrated at about 2.6 percent in the atmosphere. The energy transfer due to evaporation from the world's oceans, and even from the world's land masses, exceeds the transfer caused by worldwide energy production by several orders of magnitude, so the addition of more steam from hydrogen to the atmosphere wouldn't make much difference, Schönwiese felt. Production of steam near the ground by a future hydrogen energy technology will be of "minor significance" even if hydrogen achieves a share of "considerably more than 10 percent" of global primary energy consumption. With airplanes releasing steam at altitudes below 5 miles and at temperatures higher than 35°C, increases in water and in water-ice clouds could actually reduce the greenhouse effect, Schönwiese said. However, formation of ice clouds above those altitudes and at temperatures below 35°C due to water vapor emission of hydrogen-burning aircraft might in fact increase the greenhouse effect, based on then-current climatic models, he added.

What about nitrogen oxides, the by-products of any burning in the atmosphere? With hydrogen, chances are that the problem can be diminished if not licked. At the 1973 NASA symposium, two scientists from New York University, Antonio Ferri and Anthony Agnone, said that existing types of turbojet engines burning hydrogen would actually produce more nitrogen oxides. But, they added, the situation would change completely if different combustion schemes (such as vaporizing the hydrogen, premixing it with air, and injecting it into the combustor before burning) were introduced.[7] Ferri and Agnone concluded that in future subsonic planes the formation of nitric oxides could be much reduced by combustor designs of this type.

If hydrogen makes subsonic planes look good, its use in a future supersonic transport (SST) glows with economic promise and environmental health, although studies are still needed to determine whether and how increased water vapor (in itself harmless) might change weather dynamics in the high stratosphere.

A proposed American SST, which Boeing had planned to build after winning a design competition against Lockheed, was killed by environmentalists in the US Senate in 1971. In retrospect, this was probably a good thing. Aside from pollution concerns, the plane's marginal economics probably would have sounded its death knell rather quickly after the Arab oil embargo and the energy crisis of 1973. The French-British Concorde never became a commercial success, and the TU-144 (a Soviet SST that looked much like the Concorde) was mothballed within a few years because of technical problems.

When it comes to emissions, hydrogen-fueled SSTs would have the same advantages as subsonic hydrogen-burning airplanes. They would produce no CO_2, CO, or unburned hydrocarbons at all, and the potential exists to lick the nitrogen oxide problem with advanced combustion technologies. Sonic booms could be reduced by design and by the fact that high-speed supersonic transports would likely fly much higher than today's jets (a Mach 6 plane might cruise at 100,000 feet) with less of the sonic boom shock reaching the ground.

Quick-turnaround, high-speed, and/or very large aircraft are likely to be needed to minimize congestion in the air and around airports. The International Air Transport Association projected in its 1997 long-term forecast that the volume of international air passengers would increase from 409 million in 1996 to 948 million by 2011; air freight and charter flights are also expected to grow substantially.

In the mid 1970s G. Daniel Brewer laid out his case for a hydrogen-powered SST in studies for NASA. Generally, he wrote, the same characteristics that made liquid hydrogen superior to kerosene for subsonic planes apply to an SST—only more so. Using the Mach 2.7, 234-passenger, 4200-nautical-mile basic configuration with which Lockheed lost to Boeing in the 1971 competition, Brewer found the hydrogen-fueled version was longer but also almost one-third narrower. (Narrowness would facilitate taxiing and parking in congested airports.) The weight savings for the LH_2 version would be stupendous. At 167 tons gross weight, the plane would weigh only half as much as an equivalent Jet-A plane. It would need only one-fourth the fuel weight of the Jet-A plane to fly the same distance, and the engines would have to provide little more than half as much thrust. For identical payload, range, and speed, the LH_2 plane would have to carry 37 tons of fuel, the Jet-A plane 148 tons. Thus, the LH_2 SST would also be more energy efficient. According to Brewer, it would require 4272 Btu per seat per mile. The Jet-A design would require 6102 Btu—43 percent more.

The environmental aspects that made the hydrogen-fueled subsonic transport plane attractive also applied to the LH_2-fueled supersonic plane. Both sideline and flyover noise were expected to be lower, because, again, the plane would be considerably lighter. Sonic boom pressures would likely be lower, because of the LH_2 plane's reduced weight and smaller wing area. Further pressure reductions were believed to be possible with advanced designs.

A unique characteristic of liquid hydrogen is its ability to absorb large amounts of heat. For a plane flying at multiples of the speed of sound, liquid hydrogen's vast cooling capability becomes an important additional asset. For a Mach 6 plane zooming along at 90,000 feet, for instance, the friction of the hypersonic airstream builds up temperatures of up to 2500°F at the nose and at the leading edges of the fuselage, the wings, and the control surfaces. Ultra-cold liquid hydrogen would be funneled through carefully spaced cooling tubes through all the critical areas where heat buildup occurs; it would also cool the flight deck and the passenger compartment. A secondary liquid might be circulated through the airframe and the wings to carry heat to a heat exchanger, where it would be transferred to the liquid hydrogen to impart an extra measure of energy before combustion in the engines.

In sum, then, the verdict regarding the use of liquid hydrogen for long-range subsonic and supersonic commercial aircraft, is, in Brewer's words, "an enthusiastic yes."

Almost immediately that view ran into criticism, some of it quite severe. Some critics have argued that in terms of total cost—including the expenses of production, transportation, and storage—liquid methane may be less expensive than and therefore economically preferable to liquid hydrogen.

For reasons of safety and availability as well as for environmental reasons, Brewer dismissed liquid methane as an inferior choice. He argued that, since civil commercial aviation is a truly international business, with airplanes refueling in New York, Rome, Tokyo, Jeddah, Warsaw, and Karachi, a future synthetic aviation fuel must be equally international—locally producible, with uniform characteristics everywhere. This is no problem as long as petroleum is cheap, and at the time kerosene and other hydrocarbon fuels were available at low cost all over the world. But aircraft makers, who have to think decades ahead when planning the useful life of airplanes, must face the fact that, beginning sometime in the period 2010–2020, petroleum may cease to be cheap. Brewer contended that it was going to be necessary to develop a new generation of synthetic aviation fuels that would be available everywhere around the world.

Liquid methane would be all right for New York, Beijing, and Warsaw, since the United States, China, and Poland have plenty of coal. But what about Rome? Tokyo? Karachi? Jeddah? Italy, Japan, Pakistan, and Saudi

Arabia have no coal. However, Italy could produce hydrogen from solar power; so could Pakistan and Saudi Arabia. And Japan is looking at the idea of eventually importing hydrogen on a massive scale from such potential low-cost producers as Canada, China, and Latin America. As Brewer observed in 1976:

If international air travel is to continue to flourish and expand as projected in the face of definite prospects that some countries may be unable to obtain adequate supplies of petroleum at all times, it becomes mandatory either that all nations agree to share their petroleum fuel supplies or that they adopt an alternate fuel that can be commonly produced without hazard of control by a cartel.

Hydrogen offers many potential advantages for this application including the facts that (1) it can be manufactured from coal and water, or from water directly, using any of several processes and a wide variety of possible energy sources, and therefore can be considered to be free of the dangers of cartelization and (2) used as a fuel for aircraft it has been shown to provide significant improvements in vehicle weight, performance, and cost and to result in reduced pollution of the environment.

During the 1990s there was little progress in hydrogen-powered flight. However, there was a lot of action in trying to find a cheaper-to-operate successor to the liquid-hydrogen-and-liquid-oxygen-fueled Space Shuttle. Conceived as a low-cost utility truck of the space age, the Shuttle had turned out to be less utilitarian and much more costly to operate than its designers had anticipated.

The 1980s saw the beginning of a quest for an aerospace plane that, instead of blasting off vertically from a launch pad as the Shuttle did, would take off from a long conventional runway, go into orbit, re-enter the atmosphere, and land like a conventional plane, to be refueled with liquid hydrogen and to be ready to take off again with a shorter turnaround than that required by the Shuttle and at a fraction of the Shuttle's cost.[8] Senator Matsunaga's 1988 allusion to a "trans-atmospheric aircraft" was a reference to what Ronald Reagan had called the "Orient Express" in his February 1986 State of the Union address. ("We are going forward with research on a new Orient Express that could, by the end of the next decade, take off from Dulles Airport and accelerate up to 25 times the speed of sound, attaining low-Earth orbit or flying to Tokyo within two hours.")

By 1985 or 1986 the "Orient Express" idea had morphed into the military's "National Aerospace Plane" (NASP) project: a single-stage suborbital hydrogen-fueled airplane-rocket combination that could reach any point on the globe within a few of hours. Begun 2 months after Reagan's 1986 State of the Union address, the NASP project, a high-priority, high-visibility joint effort of NASA and the Air Force, was the largest aerospace-plane program to be initiated in the latter half of the 1980s. (Other such efforts were mounted by Britain, Germany, and the Soviet Union.)

NASP was to be used as an experimental vehicle to test high-risk technologies for Reagan's Strategic Defense Initiative, but also to ferry men and equipment to future space stations more economically than could be done with the Space Shuttle. It would take off like a conventional airplane. Hydrogen-fueled, air-breathing "scramjet" (supersonic combustion ram jet) engines would push it at speeds of up to Mach 25 to the almost-vacuum of near space, where rocket power would provide the final push needed to reach a space station.

For a while, NASP officials, anxious to find allies, courted the small international hydrogen community. For example, Robert Barthelemy, director of the NASP joint program office, brought more than a dozen NASA and Air Force scientists and NASP contractors to the 1990 World Hydrogen Energy Conference in Hawaii, where they presented papers on various aspects of NASP-derived hydrogen energy technology.

One technology investigated by the NASP team was the use of "slush hydrogen," a mixture of solidly frozen hydrogen and liquid hydrogen. With a 50-50 ratio of solid to liquid hydrogen, a 15 percent reduction in tank volume could have been achieved, thereby significantly reducing the NASP's takeoff mass. Slush hydrogen continued to be investigated in the mid 1990s in Europe and in Japan. In Europe, aerospace researchers were exploring the fuel's properties as part of the European FESTIP (Future European Space Transportation Investigation Program) effort. In Japan, scientists were looking at slush hydrogen for a more down-to-earth application: long-distance transport of hydrogen by ships.[9]

Ultimately, the NASP program was killed in the series of budget crunches in the early 1990s after expenditures of about $2 billion dollars. As far as is officially known,[10] no NASP was ever built.

In the mid 1980s, British Aerospace developed the concept of a high-speed HOTOL (horizontal takeoff and landing) craft, envisioned as a recoverable substitute for the throw-away rockets then used to place payloads into space.

Like the NASP, it was to be partially air-breathing; however, it would not have achieved the extreme speeds and altitudes of the NASP. At the end of the 1980s, Germany developed the Sänger concept. Unlike the single-stage HOTOL and NASP, the Sänger was to be a two-stage, two-vehicle machine. With a big, manned, air-breathing, hydrogen-fueled, turbo-ramjet-powered first stage, it would take off from conventional runways and accelerate almost to Mach 7, whereupon a smaller vehicle (manned or unmanned) would continue into space, propelled by a rocket engine fueled by liquid hydrogen and liquid oxygen. Neither a HOTOL nor a Sänger was ever built.

In the United States, the vigorous drive for a simpler, cheaper, hydrogen-fueled aerospace craft continued. In 1991, just about when NASP was starting its glide toward oblivion, the Strategic Defense Initiative Office (SDIO) began another program aimed at efficient, hydrogen-powered space transportation. This program, dubbed Single-Stage-to-Orbit and later renamed Single-Stage-Rocket Technology (SSRT), no longer called for NASP-type air-breathing scramjet engines of the type envisioned for NASP to get close to orbit; it called for good old reliable rocket power with a twist. Instead of throw-away rockets, the goal was to develop a rocket-powered craft that could land, be refueled quickly, and take off again. In a first competition round, SDIO selected a team led by McDonnell-Douglas Space Systems to develop a fully reusable single-stage-to-orbit launcher. In August 1993, McDonnell-Douglas flew the first of these new generations of craft, the Delta Clipper-Experimental (DC-X).

On its maiden flight, the DC-X, powered by four rockets, rose slowly from its launch pad in the New Mexico desert to about 100 feet, hovered, moved about 100 feet sideways, then slowly descended again, coming to rest on four legs protruding from its tail. In June 1994, the DC-X reached an altitude of 2600 feet in a test flight that was marred by a serious accident. During the launch, venting hydrogen was blown back by winds into a support duct, where it came into contact with the igniting engine; the ensuing explosion blew a gaping hole in the craft's metal skin as it ascended. According to *Aviation Week & Space Technology* (August 18), pieces of the skin started to fall off several seconds into the launch, when the craft had already reached an altitude of 1000 feet. Nonetheless, the DC-X landed as planned, riding its rocket plume safely to the ground.

The DC-X flew eight times; its successor, the DC-XA, flew four times. During its last flight, one of its four landing struts failed to deploy properly. The magnificent machine toppled over on its side and was destroyed. It was the end of that program.

Separately, NASA launched two competitions for a reusable launch vehicle (RLV), which eventually narrowed down to a race to build the larger, more ambitious of the two craft originally contemplated: the X-33. Lockheed Martin won in 1996 with a design for a lifting-body vehicle powered by a "linear aerospike engine" fueled by hydrogen and oxygen. Until that time, no craft of this type had actually flown except for the DC-X and the DC-XA. The X-33 is now under development. *Popular Science*, in an October 1996 article, described this high-tech flyer as looking like "a potato . . . with a '61 Cadillac tail end." Like the Space Shuttle, the X-33 will lift off vertically but will glide to a landing on a runway like an airplane. A demonstrator, it will achieve only about 60 percent of orbital speed; its proposed bigger brother, VentureStar, is supposed to be able to achieve orbit at what its developers claim will be one-tenth of the payload cost of today's launchers.

In contrast to the vigorous efforts just discussed, development of liquid-hydrogen propulsion technology for jet planes is going nowhere. Lockheed's interest in hydrogen as a fuel for jet aircraft waned with the retirement of G. Daniel Brewer in 1985 and of his boss Willis Hawkins in 1993.

Lockheed, Hawkins, and Brewer could have beaten the 1988 Tupolev hydrogen plane by about 10 years with government and industry support. As early as 1978, Lockheed began cautiously floating the idea of a small experimental cargo airline linking the United States, Europe, and the Middle East and employing four wide-body Tri-Star jets converted to liquid hydrogen. Lockheed felt that this Liquid-Hydrogen Experimental Airline Project (LEAP) would convincingly demonstrate the technical, environmental, and potential economic advantages of liquid hydrogen as a jet fuel. As was outlined at the World Hydrogen Energy Conference in 1978 in Zurich,[11] Lockheed wanted to deploy the planes on a circuit linking Pittsburgh, Frankfurt, Riyadh, and Birmingham, England. These important industrial and commercial centers already generated enough international cargo traffic to provide reasonably high load factors for the four Tri-Stars, the reasoning went. Furthermore, the four airports are close enough to existing primary energy sources (coal in the cases of Pittsburgh, Frankfurt, and Birmingham, oil and natural gas in the case of Riyadh) to permit setting up small hydrogen production and liquefaction plants at those airports.

A September 1979 International Symposium on Hydrogen in Air Transportation was organized to come up with practical ideas for LEAP. A comprehensive technology-development program was developed, but LEAP failed to win wide acceptance; indeed, *Business Week* reported that support for it ranged "from negative to lukewarm."[12] An ad hoc executive group with representatives from the United States, Canada, France, Germany, Belgium, Japan, and Saudi Arabia continued to meet and plan for more than a year afterward, but the project simply faded away.

In Europe, the story was more encouraging, but in the end the result was the same. In March 1989, a little less than a year after the Tupolev flight, the German newspaper *Welt am Sonntag* reported that the German aerospace manufacturer Messerschmitt-Boelkow-Blohm (MBB) had decided to convert an Airbus to run on liquid hydrogen, and that the plane was to be operational within 7 years. The paper even published an illustration. The plane would have a range of about 1000 nautical miles. The project would be headquartered in Hamburg—a logical venue since that city was to be an LH_2 transfer point in the Euro-Quebec Hydro-Hydrogen Pilot Project (EQHHPP).

At first MBB wouldn't confirm the report; it characterized the details as mere in-house "suggestions." However, in June 1989, at the Paris Air Show, MBB confirmed that it was "seriously" considering conversion of an Airbus to liquid hydrogen, that a working group consisting of a dozen companies was gearing up for a feasibility study, and that there were hopes for a first test flight in the mid 1990s. "We want to go at this full tilt," an MBB executive said in an interview. "We're going to be stopped only by facts that speak against it."

In May 1990, at the Hannover Air Show, the Soviet Union and West Germany—two countries in the midst of dramatic political transitions—signed a preliminary agreement to jointly develop a "Cryoplane" for civilian purposes, The original Russian LH_2 plane, the TU-155, was on display—apparently the first time that the plane was shown in the West. It had logged 15 hours flying on liquid hydrogen and 40 hours on liquid methane, according to a May 28 story in *Aviation Week and Space Technology*. "There are no major technical obstacles for the use of liquid hydrogen or liquid methane for transports," Vladimir Andreev, chief designer for the Tupolev Design Bureau, told the magazine. "Our first tests were with liquid hydrogen, and we now are using natural gas—but if we wanted to switch back to liquid hydrogen, we can do this without a problem."

The Cryoplane project brought together more than a dozen companies, including Tupolev and the engine manufacturer Kuznetsov/Samara. The basic idea was to convert either a Tupolev jet or an Airbus to run entirely on liquid hydrogen. The Germans pushed for the Airbus because it would be easier to sell in Western markets.

The most visible design change would haven been the addition of a "hump" to house the liquid-hydrogen tanks. It would add drag and degrade performance, but Airbus engineers believed it would be much preferable to either redesigning the wings (the traditional place for fuel tanks) or carrying the fuel in two tanks, one in the front and one in the rear of the fuselage, as had been proposed earlier by American designers. Airbus engineers calculated that the use of super-light liquid hydrogen as a fuel could add valuable revenue-earning capacity for cargo and/or passengers.

In 1990 the partners in the Cryoplane project hoped that a liquid-hydrogen-fueled Airbus or Tupolev TU-154 could be flying in about 5 years. That was not to be. Two years later, Airbus concluded there was no justification for building what would be a very expensive hydrogen-powered demonstrator of that size. Instead, the partners decided to focus on the development of combustion chambers, pumps, valves, tanks, and other components. Heinz Klug, director of project definition for Deutsche Airbus, believed that at least 3 years of solid basic research was needed before the partners could even begin to think about building an actual plane. "If we do well, it may fly around the year 2000," he said.

In 1994 the project took a new tack with the news that Daimler-Benz Aerospace Airbus (DASA), the new corporate entity that had absorbed MBB and its Deutsche Airbus division, was now considering conversion of a smaller commuter plane to liquid hydrogen rather than an Airbus. That plane would be a DO-328, a twin-engine turboprop plane made by Dornier, a venerable company that had been acquired earlier by Daimler-Benz. Hans-Werner Pohl, the DASA engineer in charge of the project, described the new concept at a 1995 hydrogen conference in Montreal. The main reason for the switch was that converting a smaller plane would cost less than converting a big jet transport.

Work on the plane, expected to cost about DM 60 million ($38.7 million at the time), was scheduled to start in January 1997 despite some uncertainties about financing and lack of a final signoff on the project on part of top DASA management. Money was a major concern. Pohl said at the 1996 World Hydrogen Energy Conference in Stuttgart that the basic rule was "to keep costs as low as possible." This meant using off-the-shelf items rather than optimized equipment wherever possible—for example, a conventional aluminum fuel tank with conventional insulation, and a liquid-hydrogen coupling and nozzle already fabricated by Messer-Griesheim for road vehicles and suitable, after modifications, for aircraft use.

The project was made potentially more attractive in commercial terms by a change of engines. In the summer of 1996, 80 percent of Dornier's shares were acquired from Daimler-Benz Aerospace by the Fairchild Corporation. To make the plane more interesting to commuter airlines, Fairchild opted to switch to jets from the original turboprops. DASA, which kept one plane for the LH₂ project, decided that it made sense to go with jets for the hydrogen version as well but to leave other aspects of the plane—wings, control surfaces, etc.—essentially intact. Two engine makers, Pratt & Whitney Canada and Allied Signal, had initially expressed interest in supplying a power plant.

DASA also began to look for solutions to the NOx problem associated even with the burning of ultra-clean hydrogen. At the 1996 World Hydrogen Energy Conference, a joint paper by DASA, Pratt & Whitney Canada, Hydro-Quebec, and the German aerospace agency DLR described the outcome of a four-year experimental program to establish how much potential existed for NOx reduction and to provide guidelines for the design of low-NOx combustors. The researcher examined the potential of the two concepts regarded as most promising: the "premixed perforated plate" and the "high shear swirl" injectors. Both relied on rapid mixing of fuel and air, plus rapid burning and quenching of the gases, to minimize NOx. "Based on the test results from this program, practical low-NOx hydrogen gas turbine combustors can be achieved," the authors concluded. In another paper, two scientists from Aachen Technical University described their separate efforts to design low-NOx hydrogen aircraft gas turbines for the Cryoplane. Günter Dahl and Friedemann Suttrop reported that they had developed a "micro-mix diffusive combustion" system that achieved "safe diffusive hydrogen combustion with substantial reductions of the NOx level."

In the end, all this work and all these plans came to naught. In early 1999, management decided to put the project on hold indefinitely because of financing problems: a shortfall of DM 10–15 million ($5.3–8.8 million).

In the 1990s, DASA, Tupolev, and their allies were almost the only aircraft producers taking any interest in hydrogen as an aviation fuel. Others showed no visible interest—least of all Boeing, the giant among the world's aircraft manufacturers. A Boeing contribution to the 1993 volume *Transportation and Global Climate Change*, published by the American Council for an Energy-Efficient Economy, reiterated the company's long-held position that hydrogen was something to be considered only for the distant future. The paper, "Characteristics of future aviation fuels," authored by O. J. Hadaller and A. M. Momenthy, acknowledged that there was "an increasing number of environmental concerns, such as global warming, driving the search for a replacement of petroleum-based fuels." "However," it continued, "studies indicate that the currently used aviation fuel is as likely to satisfy these concerns as the few alternative fuels that are suitable for use in aircraft. Hydrogen . . . will become economically acceptable only after the world has exhausted its fossil fuel resources or a low-cost, abundantly available source of electric power, such as nuclear fusion, is developed. . . . Improving efficiency will be the principal way to lessen the impact of aircraft on the environment until a technically and economically practical non-fossil based fuel is discovered." An accompanying chart indicated that world passenger aircraft efficiency, expressed as revenue passenger miles per gallon, had roughly doubled between the early

1970s and the mid 1990s, from somewhat below 20 miles to just about 40 miles.

At NASA, on the other hand, there is interest in hydrogen on the part of some officials. NASA participates in global efforts to assess the effects of airplane emissions on the atmosphere and is attempting to come up with ideas for reducing them. In the eyes of these officials, hydrogen stands out as a prime candidate.

In the mid 1990s, Richard Niedzwiecki of NASA's Lewis Research Center headed a NASA section that worked principally with gas turbines. His scientists investigated combustion design and fuels and their emissions as part of atmospheric research. Briefing the Department of Energy's Hydrogen Technical Advisory Panel (HTAP) in November 1996, Niedzwiecki said that European countries were in the forefront of assessing aviation's impact on the atmosphere. Niedzwiecki pointed out that aviation emissions were responsible for about 3 percent of the world's CO_2 output. Some forecasts "are talking about 10 percent by the year 2050," he said, but "I've seen (estimates of) 10 percent by 2015." Niedzwiecki explained that world travel was expected to grow 5 percent a year through the year 2010, equaling a potential travel market of $800 billion. Part of the equation is the market for some 3200 jet engines, and these engines will have to be much less polluting. The trend is clear and ominous: With a million high-quality, high-paying jobs and sales of $100 billion (1991) at stake, aeronautics plays a big role in the US economy, Niedzwiecki said. The problem is that others want to move in on these high-technology markets: "We were the only game in town maybe a generation ago when it came to selling advanced aircraft to the world," Niedzwiecki noted. "Now we have consortiums in Europe"—Airbus Industries—"which already have a 30 percent share of the international market for large civil transport aircraft and which already use environmental compatibility as a marketing tool. Also, there are suggestions coming down the pike to impose a CO_2 budget on all transportation systems after the turn of the century, including aircraft. This could very well mean that aircraft CO_2 is frozen at a certain level—let's say, a 1995 level." "You can make up some of that in terms of more efficient power plants," he went on. "But there is a question as to how far you can go. In essence, you would freeze growth in that industry doing that. And that's of tremendous, grave concern in aeronautics as it should be in ground transportation."

Looking at all possible fuel alternatives that don't produce CO_2, Niedzwiecki said: ". . . the only thing I can come up with that is viable as a fuel is hydrogen. . . . Hydrogen is an excellent fuel for many applications, including aircraft." On the question of whether to tap into fossil fuels in the transition or whether to go directly to splitting water as a hydrogen source, Niedzwiecki left no doubt where he stood: "Environmentally acceptable manufacture . . . is the most important thing of all. I've heard here today that we could make hydrogen out of different (hydrocarbon) fuels, and then take the hydrogen and use it. . . . I might suggest an atmospheric scientist would wince at that. The general conclusion is, CO_2 is CO_2. . . . It doesn't matter if you produce it on the ground in Timbuktu or you produce it at 90,000 feet right over the White House." Either way, CO_2 would be injected into the atmosphere. Niedzwiecki continued: "I would strongly suggest that programs in the area of making hydrogen from non-hydrocarbon fuels would certainly be a priority of mine." Combusting hydrogen in a jet engine would produce almost 3 times the amount of water in the atmosphere than burning kerosene-type aviation fuel to produce comparable thrust, says Cecil Marek, a technical advisor to a NASA team at the Glenn Research Center that has again started looking at hydrogen fuel for airplanes. But kerosene produces 25 percent more of other emissions, and "water washes out of the atmosphere a lot faster than carbon dioxide which stays up for years," Marek explains. Still, injecting more water into the atmosphere creates more uncertainties about the formation of high-altitude cirrus clouds and their effects, all of which would need more study, according to Niedzwiecki. But "any atmospheric scientist you talk to would say that's a thoroughly acceptable trade at the present time."

Niedzwiecki acknowledged that the issue of high hydrogen costs versus current low jet fuel costs is going to be fiendishly difficult to deal with. He explained that it costs about 8.9–12.5 cents to process a gallon of jet fuel today; the rest is profit for various other functions such as transport, storage, and distribution. For now at least, many people in the industry will resist switching to a revolutionary new fuel such as hydrogen because of fear of escalating costs and because they want to stay in business, he said. But "if it's an environmental issue, and if it leads towards the slowing down or the elimination of growth in the aircraft industry, that cost will not be a discriminator (for) people that work in those fields, and I think [it] probably would be gladly accepted."

Lockheed veteran G. Daniel Brewer also attended that meeting, and he briefed the panel on his 1970s findings. Extrapolating from FAA statistics and EPA data, Brewer concluded that the more than 2000 takeoff and landing cycles at the four major airports in the Los Angeles area generate more than 95,000 pounds of NOx and more than 5000 pounds of particulates from Jet-A fuel every day in the Los Angeles basin, as well as CO, CO_2, oxides of sulfur, and unburned hydrocarbons.

Brewer, who in 12 years as Lockheed's hydrogen manager produced "several million dollars' worth of studies for subsonic, supersonic and hypersonic" airplanes, said that a technology development program should be launched that would draw on the planning work done at Lockheed and elsewhere. But the prospects don't look good, he acknowledged: "Apparently, the government agencies involved in aircraft development require evidence of industry interest and support. That's not available because the aircraft industry and the air transport industry aren't convinced that there will be fuel available at the time the aircraft are available. So, obviously, they won't start spending money to develop a new type of aircraft." The point, said Brewer, "is that in spite of these advantages, the overall support for use of hydrogen by our government—that is, NASA, FAA and all the other government agencies that are involved in this choice—is nil."

In the summer of 1999, Brewer was still pessimistic—perhaps more so, since DASA had dropped its modest commuter plane project, and since no change was in sight. NASA's ongoing High Speed Civil Transport (HSCT) program was focused on kerosene-burning aircraft that might fly around the middle of the twenty-first century. Brewer mused: "Although NASA officials claim they have not closed the door to hydrogen, one wonders. Is this a farsighted program?"

A year later, things began to look up again in a modest way. In the summer of 2000, hydrogen activists within DaimlerChrysler Aerospace Airbus

(DASA) launched a two-year systems analysis of liquid-hydrogen-fueled aircraft in a Europe-wide context. With the support of the European Commission, and in collaboration with 33 partners from industry and research institutions in 11 European countries, the new study once again began to look at all aspects of hydrogen as aviation fuel, including feasibility, safety, environmental compatibility, and economic viability. The new study was intended in part as a learning exercise for the various players and in part to build support for the idea among governments, in industry, and in academia, the obvious ultimate goal being to actually build an operating hydrogen airplane.[13]

In the United States, NASA had also begun once more to look at hydrogen as fuel for airplanes. In the autumn of 1999, a team of seven scientists at the Glenn Research Center in Cleveland launched a 3-year program budgeted at about $7 million to explore prospects for zero-emitting civil airplanes that would use hydrogen instead of kerosene or aviation gasoline. The "Zero CO_2 Emissions Technology Project" investigated the use of hydrogen as a fuel for jet engines and aviation gas turbines, but also as a fuel for ultra-lightweight fuel cell systems that might be used to power small (four to six passengers) propeller-driven airplanes, and maybe even larger ones. The project had its philosophical foundation in NASA's stated goal to reduce nitrogen oxide emissions of future aircraft by a factor of 3 within 10 years, and to reduce carbon dioxide emissions by 25 percent within 10 years and by 50 percent within 25 years.[14]

The NASA team was considering both PEM and solid oxide fuel cells as possible aircraft power sources, provided they could be made small and powerful enough: "We need to know whether a Cessna 172 class aircraft . . . needs fuel cell energy density 2 or 5 times that of the current state of the art," team leader David Ercegovic said in an interview.[15] Also to be determined is whether lightweight polymer composite fuel tanks to carry liquid hydrogen can be developed. Still, said Ercegovic, "I am willing to say that a Cessna 172 class aircraft testbed [operating] on hydrogen and a fuel cell is not out of the question in the next five to ten years."

Advanced Technologies, Inc., of Worcester, Massachusetts, a small company specializing in ultra-high-energy batteries and auxiliary starting systems for aircraft, said in 2000 that it was working on experimental planes powered by ultracapacitors or batteries and that it was about to start converting two lightweight, low-drag motorgliders (an Austrian Diamond Katana and a French all-carbon BanBi) to fuel cell power, with an initial goal of building a plane that could take off, fly at least 100 miles, and land safely. Future goals include a fuel cell plane with a range of 300 miles or more.[16]

Maybe there will be silent, electric ZEV airplanes in our future.

9

Hydrogen as Utility Gas: The Invisible Flame

In the late 1960s, the Institute of Gas Technology (IGT) exhibited a "Home for Tomorrow" said to be powered by "reformed natural gas," a "new super active form of natural gas." An illustrated four-page color brochure described new types of illumination, portable appliances, flameless wall-panel heaters, total climate control, and electricity generated in the house by means of fuel cells. The "super active form of natural gas" was simply a hydrogen-rich mixture, similar to the manufactured town gas or coal gas of the late nineteenth and early twentieth centuries. Whereas town gas typically had up to 50 percent hydrogen, "reformed natural gas" would have contained about 80 percent hydrogen, 20 percent CO_2, and less than 0.5 percent CO.

The IGT brochure envisioned that straight natural gas would be piped to the home. Except for the portion that would be burned directly (e.g., for space heating), the gas would be piped into a compact "reformer" that, using heat and steam with the help of catalysts, would convert natural gas into "reformed natural gas"—mostly hydrogen. According to a 1970 article in the industry magazine *Appliance Engineer*, such a use of hydrogen would allow the use of catalytic burners on kitchen stoves. Unlike conventional burners, which burn with a blue flame, these burners' flames would be invisible. They would transfer more usable heat. Temperature ranges would be much wider and could be adjusted much more finely, and the heat would be distributed much more evenly over the burner's surface. A catalytic burner, said the article, "is self-starting, requiring no pilot, glow coil or spark ignition." Ignition would take place when the hydrogen gas contacted the catalyst (typically a very thin coating of platinum) and the burner began to give off heat. The by-product would be water vapor. Because of the fine-tuning capability and the wide range of heat available, "water can be boiled very rapidly or the most delicate sauce kept barely warm," the *Appliance Engineer* article said. A warming tray could be made of wood because the temperature could be limited to a maximum 250°F, below wood's burning or scorching temperature. Room heaters could operate at low temperatures and could be hung on walls and covered with synthetic materials. "Reformed natural" gas was also to be the power source for candoluminescence, a novel illumination process in which light emitted by molecules, ions, and atoms would be stimulated by a flame. "The physical causes of candoluminescence are still unknown, but we do know that this phenomenon provides new product possibilities," the article said.

Though nothing commercially saleable came of these "Home of Tomorrow" pipe dreams, the research that was behind them provided a valuable first look at many of the technical problems that will have to be solved if hydrogen is ever to come into wide use as a fuel for the home. Many of these early conclusions are still relevant today.

The Institute of Gas Technology also began developing high-temperature catalytic gas burners. At the 1976 World Hydrogen Energy Conference, held in Miami Beach, three IGT researchers, Jon Pangborn, Maurice Scott, and John Sharer, presented a paper on the technical problems of burning hydrogen in modified conventional burners and in advanced catalytic burners. Pangborn, Scott, and Sharer said that hydrogen could not be directly substituted for natural gas in domestic and commercial appliances and that burners would have to be modified, a prospect they did not regard as daunting. "Similar equipment modifications were necessary when natural gas was substituted for manufactured gas several decades ago in the United States," they explained.

Conventional burners suffer from flashback when burning hydrogen. Although the flame will not travel back into the pipe beyond the metering orifice (because beyond that point there is no air or oxygen to sustain the flame), a flame traveling back to the orifice may damage the burner head. The problem can be tackled either by increasing the pressure of the gas and using a smaller burner port or by reducing or eliminating the amount of so-called primary air—air mixed into the gas stream near the metering gas orifice.

Though hydrogen does not produce any carbon monoxide or any unburned hydrocarbons, an open-flame hydrogen burner would produce about 30 percent more nitrogen oxide emissions (due to the flame's reaction with nitrogen in the ambient air) than a comparable natural gas burner. Pangborn, Scott, and Sharer felt that most gas appliances could be made compatible with hydrogen by increasing the pressure drop and the flow rate at the user's end and by redesigning the burner.

A few experimental hydrogen-powered houses have been built by dedicated individuals and institutions. One of the earliest hydrogen home conversions was undertaken by Olof Tegström in the mid 1980s. Primary power for Tegström's 1334-square-foot two-story home in Haernoesand, on Sweden's east-central coast, was provided by a 72-foot-high windmill with a three-blade propeller 49 feet in diameter. With two generators (one for high and one for low wind speeds), the windmill generated about 40,000 kilowatt-hours per year. On average, 1 kW was used to power domestic appliances, 5 kW for hydrogen generation. Hot water was stored in a 5-cubic-meter tank for house heating. Tegström also sold some electricity back to the grid. A modified kitchen stove with a stainless steel grid made the hydrogen flame visible, generated almost no nitric oxides, and provided humidification. In all, Tegström's highly insulated house required about 20,000 kWh/year. Hydrogen was generated in an 84-volt SPE electrolyzer that, at 50 amperes, produced 1 standard cubic meter of hydrogen per hour from 4 liters of tap water. After drying, the hydrogen was stored in a hydride tank, apparently made of a fairly standard iron-titanium alloy.

Tegström also converted his Saab 900 automobile to hydrogen power. He estimated that converting both the house and the car had cost the equivalent of about $139,000 (at average 1998 exchange rates). But "these outlays should be seen as a down payment for 20 years of energy costs," he said.

In the late 1980s, Walt Pyle began converting his 1800-square-foot two-story house in Richmond, California (near the northern end of San Francisco Bay) to photovoltaic and hydrogen energy. Pyle had started on his quest for clean power for his house in the mid 1970s by investigating photovoltaics and adding energy-efficiency improvements. Pyle (once a staff engineer at a major oil company and later a solar-energy entrepreneur) is still at it, and he has documented the ongoing process in a series of articles in *Home Power Magazine*.[1]

For primary power, Pyle relies principally on 52 photovoltaic panels producing about 1500 watts peak. (For short-term storage of electricity produced by these panels, Pyle uses two strings of six industrial-type batteries with a total capacity of about 1600 ampere-hours. "The biggest problem is maintenance of the batteries," he says. "If you deep cycle them, they don't last very long.") He has installed a small 1-kW electrolyzer made and sold by Hydrogen Wind, Inc., of Lineville, Iowa. He stores the hydrogen and oxygen in three converted medium-pressure tanks originally rated at about 250 psig for liquefied propane gas but derated to 60–125 psig for hydrogen. He has also built an experimental metal hydride storage system, using mostly iron and titanium, plus an AB2 alloy (zirconium, vanadium, nickel, chromium and manganese) made by GfE of Nürnberg. Each of the three tanks holds about 0.45 cubic meters of hydrogen.

For cooking, Pyle has converted a conventional gas stove to run on hydrogen, modifying the burners to avoid any mixing with air before the hydrogen reaches the burner ports to avoid flashback. Pyle has also converted a standard propane-powered outdoor barbecue grill to hydrogen operation, and he has looked at catalytic combustion with the aid of a stainless steel mesh. H-Ion Solar has commercialized a hydrogen cooking grill under the trade name Hydro-que.

For home heating, Pyle has converted four conventional catalytic multifuel wall-mounted space heaters, originally designed to burn natural gas and to be used in campers and recreational vehicles, to hydrogen or natural gas operation. These heaters produce infrared radiation that warms an object or a person without heating the surrounding air. (Today, about three-fourths of the house is insulated with the equivalent of about 6 inches of fiberglass.)

Pyle stays off both the gas and the electricity grids for about 9 months out of the year. From mid November through February, he admits, "we need help" from the gas grid to heat the house. Pyle doesn't have a seasonal hydrogen-storage system with enough storage capacity. He also occasionally taps into the electric grid "to let the batteries recover." Pyle estimates that he has spent about $30,000 to retrofit the house.

In 1990, the architect Markus Friedli switched his 2367-square-foot home in the small Swiss town of Zollbrueck to solar and hydrogen energy and took it off the local electricity grid.[2] Friedli's 4-year conversion program, assisted financially by the canton (province) of Berne, made use of commercially available components, including photovoltaic panels made by Siemens, an alkaline electrolyzer made by VCST Hydrogen Systems of Belgium, and a hydride vessel made by Japan Metals & Chemicals. Also installed were a hydrogen purifier and a compressor. Friedli reported that he was using hydrogen to cook and to run his washing machine. His hydrogen pipes were welded or screw-joined to gas-proof standards and were made of embrittlement-proof high-grade steel. Rooms containing hydrogen machinery were equipped with sensors able to detect hydrogen concentrations as low as 0.14 percent, set off alarms, and shut off the main hydrogen pipe. The total cost of the conversion at the time was said to be the equivalent of about $216,000.

The most ambitious hydrogen house yet was in Freiburg, Germany. It was designed and built by the Fraunhofer Institute for Solar Energy Systems. Planning for the $1.5 million, 1566-square-foot, two-story experimental house (about $1 million for the sophisticated energy-efficient structure, the rest for solar and hydrogen energy technology) began in 1987. Hydrogen was used for long-term energy storage. Through the use of advanced foamed glass, transparent heat dam materials, and insulation made of reprocessed waste paper, heat losses were cut by more than 70 percent relative to conventional German buildings. The house went "on sun" in 1990.

Visually, this grid-independent and furnaceless house was distinguished by a glass-covered curved facade facing south and by the absence of a chimney. Mounted on the roof were 40 square meters of photovoltaic cells plus thermal solar collectors. A 2-kW electrolyzer in the basement generated hydrogen for long-term storage for the winter. (A bank of lead-acid batteries storing about 20 kWh was used for short-term solar electricity storage.) Hydrogen was re-converted into electricity by a fuel cell, and the specially designed four-burner catalytic stove cooked with hydrogen.[3] Hydrogen was stored as a gas in an above-ground tank alongside the house; it could hold the equivalent of 1400 kWh. Oxygen was stored in an underground tank.

"This is not a prototype that everybody will be able to build tomorrow," said Adolf Goetzberger, the Fraunhofer Institute's director and also the president of the International Solar Energy Society, at the dedication in late

1990. But, he added, "it is designed to show that the goal of energy independence is basically achievable." (At the end of 1995, the project was stopped. Six months later, the house was connected to the grid and turned into an office building.)

The use of hydrogen for utility applications came into the picture in a small experimental way with the development of new devices to help manage power fluctuations in the electrical grid. One such device, developed by the German aerospace agency DLR in the late 1980s, was a so-called stand-by hydrogen/oxygen spinning reserve steam generator that could produce steam nearly instantly for use in conventional steam generators during periods of sharply rising electricity demand. The device consisted essentially of a much-modified rocket combustion chamber that burned hydrogen and oxygen and generated steam at efficiencies of about 99 percent over a wide range of power, temperature, and pressure with extremely short startup times, reaching maximum output in less than a second. A joint paper presented by DLR and DASA at the 1996 World Hydrogen Energy Conference in Stuttgart said that work on the device was continuing and more funding was needed, but it was pretty clear that this was just about the end of the road.

Hydrogen production and grid load management are the complementary capabilities of the advanced high-pressure electrolyzer developed by GHW. The company began testing the initial 100-kW prototype alkaline filter-press type machine in 1992. In late 1997, the company announced it was contributing a 500-kW, 300-bar unit to a demonstration facility at Munich's airport that would integrate hydrogen production, various modes of hydrogen storage, and a robotic hydrogen refueling station to service hydrogen-powered ramp vehicles. The facility was inaugurated two years later.[4]

In the planning stage is another 500-kW electrolyzer that is to furnish the fuel for a hydrogen-powered bus in the German city of Karlsruhe. Unlike conventional electrolyzers, which have to operate at a fairly constant rate, the GHW machine adjusts quickly and automatically to wide swings in the power supply—from 15 percent to 120 percent of rated capacity. It permits power plant operators to operate at close to 100 percent of rated capacity without fears of mismatch and frequency fluctuations (which cause electric clocks to show the wrong time and tape recorders to run at the wrong speed); the electrolyzer would soak up excess power, producing hydrogen at 80–87 percent efficiency. Because of its rapid response and flexibility, the device is regarded as close to ideal for hydrogen production from intermittent sources such as wind and solar.

In return for load leveling, the electrolyzer's operators are likely to demand—and get—rock-bottom power prices from German utilities that will enable them to produce hydrogen inexpensively. According to GHW, retail prices in Germany averaged the equivalent of 14.7–18.3¢/kWh in early 1994, going as high as 51¢/kWh in some locales. At the time, GHW believed that rates as low as 2¢/kWh for hydrogen production would be possible with such a tradeoff.

The only other technology to have transcended "home of tomorrow" fantasies is the fuel cell, development of which shifted into high gear—for transportation, for power generation, and for the home—in the mid 1990s. Typically fueled by natural gas, which is first converted to hydrogen, 200-kW fuel cells suitable for utilities, office buildings, hospitals, and similar facilities are already available commercially from the International Fuel Cells Corporation of South Windsor, Connecticut. Other firms, such as the FuelCell Energy Corporation (née Energy Research Corporation) of Danbury, Connecticut, are working on megawatt-size molten carbonate fuel cell units capable of using natural gas directly, without preliminary reforming. Canada's Ballard Power Systems, generally acknowledged to be the international leader in fuel cells for transportation, is racing to extend its expertise in PEM fuel cell technology to stationary and utility applications. And the Siemens-Westinghouse Power Corporation in Orlando has several solid oxide fuel cell projects in the works, including a 300-kW combined cycle unit in cooperation with RWE (a major German utility), a 1-megawatt combined plant with the German utility EnBW, a 250-kW combined-cycle unit at the National Fuel Cell Research Center at the University of California at Irvine, and one near Bergen, Norway that employs CO_2 sequestration technology developed by Shell Hydrogen.[5]

At the end of the 1990s, with an eye toward deregulation of the utility market and the possible breakup of large centralized electric utilities, many companies[6] in North America, Europe, and Japan are competing to develop small stand-alone fuel cell systems and components, usually powered by natural gas. A partial list includes Avista Labs, H Power Corporation, Plug

Power, Manhattan Scientifics, Nuvera, Delphi Northwest Power, DAIS-Analytic, ElectroChem, DCH Technology, Thermo-Electric, ZeTek, De Nora, Sulzer, Fuji Electric, and Matsushita. Here are few snapshots of some the major players in this constantly changing field.

International Fuel Cells   In mid 1997, International Fuel Cells announced that it had won 185 orders for its 200-kW PC 25 fuel cell plants, built by its ONSI subsidiary, worth $111 million, demand stimulated in part by not-all-that-generous subsidies from the US government that will help under-write only about 75 of these orders. For the lucky ones, it will reduce the price of the plant from an average of about $600,000 to $400,000. On top of these, IFC said it had received Letters of Intent and other expressions of commercial interest for an additional 235 plants. By late 1999, ONSI fuel cell power plants had racked up 3 million hours worldwide since the first unit was installed in 1992, and the total number of plants installed around the globe stood at 200.

An ONSI PC 25 fuel cell plant operating on landfill methane began tests in mid 1996 in Groton, Connecticut in a partnership with the Northeast Utilities power pool, the Environmental Protection Agency, and IFC as a followup to an earlier six-month test at a Los Angeles landfill. The output of the fuel cell is lower than normal because the energy content of landfill-generated methane is only about half that of natural gas. Two commercial versions were installed in California in late 1998, another one by a German utility in Cologne.

A liquid-hydrogen-fired variant of the PC 25—the only one in the world so far—was installed as an experiment by two utilities, HEW and GEW, in Hamburg. It began operations in mid 1997. It was converted by ONSI's European licensee, CLC Ansaldo of Genoa. If CLC and ONSI have their way, there may be more in the future: an early CLC survey of by-product hydrogen available in Europe from various chemical operations has turned up the equivalent of some 60 MW of power in Italy, Germany, Switzerland, and the Netherlands.

FuelCell Energy Corporation   The largest yet molten carbonate fuel cell power plant, a 2-MW unit designed and built by the company under its previous name, Energy Research Corporation, began churning out electricity for the Santa Clara, California utility in the spring of 1996 in a joint demonstration project with several other utilities, the US Department of Energy, and the Electric Power Research Institute. The test terminated a little less than a year later, somewhat ahead of schedule, because of funding bottlenecks caused by unanticipated testing requirements and unexpected problems with some materials used in the plant's construction. The facility operated for about 3600 hours, a little less than half of what had been planned. ERC was considering another demonstration with a bigger plant of perhaps up to 3 MW.

An advanced variant of the molten carbonate technology licensed from ERC, dubbed a "hot module," started testing in the late summer of 1997 in Germany. Developed by a Danish-German consortium called ARGE DFC, the project had as a goal to simplify and integrate the so-called balance-of-plant equipment of such a fuel cell system into one common unit. This was important, said a consortium executive at the 1996 Fuel Cell Seminar in Orlando: while in the past innovation had concentrated mostly on fuel cell stacks, the other components had not advanced as much and were still relatively conventional. A first demonstrator plant went on line in Bielefeld, Germany, in the autumn of 1999.

Ballard Generation Systems   In late 1996, this subsidiary of Ballard Power Systems teamed with a large international power generation and distribution company, GPU International, Inc., of Parsipanny, New Jersey, to manufacture 250-kW stationary PEM-fuel-cell plants, which the company had begun to develop two years earlier. GPU invested $23.25 million in the joint venture, and the target date for first commercial sales was 2000. In late 1997, France's GEC ALSTHOM joined the group to build and sell stationary PEM fuel cells in Europe. Plans for Europe's first PEM Ballard-type fuel cell power plant, to be built in Berlin, were announced by five utilities (four German, one French) in August 1998.

MC Power   The world's largest molten carbonate fuel cell cogeneration plant, a 250-kW facility built by MC Power, began producing power at the US Navy's Miramar Naval Air Station in the spring of 1997. With cogeneration factored in, the company said, the plant has an overall efficiency of close to 80 percent. Unfortunately, the company folded in early 2000 after a year of uncertainty; in the accelerating race toward commercialization and intensifying competition, MC Power was unable to raise new funds for research and development.[7]

Other companies working on planar solid oxide fuel cell power plants include Dornier as part of the Daimler-Benz group, and, in Japan, Tonen, Sanyo Electric, Murata, and Mitsubishi.

In developed Western countries, little thought is given to the idea of reintroducing hydrogen as a component of town gas (the mix of carbon monoxide and hydrogen made from coal that fueled much urban development around the world in the first half of the twentieth century). Some energy strategists have argued that clean natural gas would be even cleaner if hydrogen were blended into urban natural gas supplies. The German consulting firm Ludwig-Boelkow-Systemtechnik floated that idea in the mid 1990s, proposing to stretch natural gas in some parts of Munich's grid with hydrogen, but apparently found no takers.

One of the last places where town gas was replaced by natural gas was West Berlin, where the change was made in the waning days of the Cold War, 15–20 years after most other parts of Germany—including East Germany—had already switched. During the Cold War, building a natural gas pipeline from West Germany to Berlin across East German territory was politically impossible. Until 1980 or so, town gas was made from coal brought in by train and canal barge from West German coal mines; after that, town gas was produced from gasoline shipped by rail or barge. The availability of town gas in Berlin facilitated the selection of that city as a showcase for the experiment of operating 10 Daimler-Benz vehicles on hydrogen in a 4-year fleet test between 1984 and 1988.

For China, cleaning up coal by means of gasification may be an economically and an ecologically acceptable strategy. China, the world's largest producer and user of coal, produces more than a billion metric tons of coal per year, which fill about 75 percent of commercial energy requirements. Coal-based energy is used in two major forms: as electricity and as town gas. The country's electric production capacity of 199 GWe (1996) was planned to grow to 320 GWe by the year 2000, 640 GWe by 2020, and 1800 GWe by 2050, according to a paper presented in October 1996 in

Beijing at a clean fossil energy seminar by Francis Lau, IGT's managing director for process development and engineering. Coal accounts for about 80 percent of China's generating capacity, a figure planned to drop to 50 percent by 2050.

Until now, town gas has played a relatively small role in China, but its importance is slated to rise considerably. According to Lau, at the end of 1993 China produced enough town gas to serve about 95 million people—about 10 percent of the population. Coal contributed 13 billion cubic meters per year of manufactured gas, but only about 3.5 billion cubic meters were for domestic use (by about 25 million people). In the aggregate, other sources of town gas were much more important. Town gas made from liquefied petroleum gas—about 2.3 million tons/year—was consumed by 57.7 million Chinese, and natural gas contributed about 1.4 million cubic meters of town gas for about 11.8 million people.

In the 1996–2000 Five-Year Development Plan, China's central planners put more emphasis on town gas development in small and medium-size cities, hoping to reach 60 percent utilization by 2010. But with a shortage of LPG and natural gas, reported Lau, the emphasis will have to be on coal as a raw energy source to meet these ambitious goals. Thus, gasification, with prodigious amounts of hydrogen produced in the process, appears to be the way to go for China's energy planners.

Beginning in the early 1980s, China began to look at various Western coal gasification technologies for ways to make more efficient use of its coal reserves and to help clean up the environment under the "trigeneration" concept. The basic idea was to produce many products from coal, including town gas, chemicals, and electricity. With double-digit economic growth rates and a consequently skyrocketing demand for energy, the Chinese government sees these as complementary issues, both of which require urgent attention.

One such process selected was the Institute of Gas Technology's U-GAS process, for which China acquired a license to build the world's first commercial plant (in Wujin). The basic job was to come up with a cost-competitive way of producing, per day, 1.7 million normal cubic meters (Nm^3) of town gas, 200,000 tons of methanol, and 7 megawatts of electricity, according to Lau. The government approved the project in 1991, ground was broken in March 1992, and the first phase of the project was completed in December 1994. Gas production started in the spring of 1995. The plant converts about 800 tons per day of low-ash, low-sulfur bituminous coal into 3 million normal cubic meters of low-heating-value industrial fuel gas, which includes about 15 percent hydrogen.

Usually, a gaseous fuel such as hydrogen is distributed by pipeline. A basic question that is still unresolved is whether hydrogen could be carried in existing natural gas pipeline networks—something that many researchers seem almost to have taken for granted at one time. However, Pangborn, Scott, and Sharer, in the aforementioned 1976 report, said that before large stretches of pipeline could be converted to carry hydrogen a great deal of research was required. "Any statement that hydrogen or hydrogen-rich gases can be adequately and safely delivered to the customer by using the in-place natural gas distribution system is a presumption," they said. There was not a single natural gas system but "a tremendous diversity of pipes and fittings." Early low-pressure manufactured-gas systems were made predominantly of cast iron. Since then, ductile and wrought iron mains have been added, followed by steel mains and services lines to the individual customer with the advent of higher-pressure natural gas. Utilities overwhelmingly use plastic pipes for new and replacement mains and services lines, the authors said. Then there are materials such as brass (used in valves), natural and synthetic rubber (used in mechanical joint seals and meter diaphragms), lead and jute (used as sealer materials), and cast aluminum (used in meter housings and regulator parts). It will have to be verified that these materials can be used safely with hydrogen.

Leakage, which some researchers considered a big challenge, was not that problematic, Pangborn et al. asserted. In terms of volume, hydrogen loss is about 2.5 times the rate of natural gas, but the energy loss is about the same as with natural gas. Plastic pipes are probably more permeable, but still the researchers felt that the amount of hydrogen that would seep through the plastic would be "insignificant."

Metallurgical embrittlement, a problem considered serious by others, was not exactly dismissed by Pangborn et al.; however, they did not regard it as prohibitive: "These modes of hydrogen attack have been observed to occur under pressure and temperature conditions far more extreme than those of a gas distribution system. . . . These metallurgical effects would not be expected to occur in distribution equipment used for hydrogen service because of the operating pressures and ambient temperatures."

Pangborn et al. made no reference to high pressures, deemed essential by other researchers for efficient energy transport; they merely assumed that hydrogen gas would flow faster and at somewhat higher pressures. If existing pipeline materials were found to be safe for hydrogen, they suggested, then if the flow rate were simply increased to about 2.8 times the rate of natural gas, but at standard, safe pipeline pressures, existing lines could deliver as much as 85 percent of the energy transported nowadays by natural gas to the end user.

This was not to say that hydrogen is easy to use in every respect. Hydrogen does create greater problems than natural gas when new lines have to be installed or when old lines have to be repaired—during welding, for instance. Also, because hydrogen burns in such a wide range of air mixture ratios (from 4 to 74 percent hydrogen), new lines must be carefully purged with an inert gas—"an area of serious concern."

If hydrogen is ever to replace natural gas as a utility fuel, very large quantities obviously will have to be stored somewhere. Storage, to maintain a buffer for seasonal, daily, and hourly swings in demand, is essential with any system for the transmission of a gas. Storage facilities even out the ups and downs of demand, including temporary interruptions and breakdowns, and still permit steady, maximum-efficiency production.

It has been suggested that huge amounts of hydrogen could be stored underground in exhausted natural gas fields, in natural or man-made caverns, or in aquifers (porous rock formations that contain water). The natural gas industry has long been using depleted gas and oil fields to store huge amounts of natural gas. Aquifers are similar to natural gas and oil fields in that they are porous geological formations, but without the fossil-fuel or natural gas content. Many of them feature a "caprock" formation, a layer on top of the formation that is usually saturated with water. This layer acts as a seal to keep the gas from leaking out; it works for both natural gas and the lighter hydrogen.

Although hydrogen is more "leaky" than natural gas, many experts feel that this has no bearing on storage, even though the materials used in injection and withdrawal equipment may pose some problems. Also, because gaseous hydrogen's heating value is lower than that of natural gas, facilities for storing it will have to be bigger and more expensive than those used for storing natural gas. Stored under pressure, hydrogen will require proportionately more volume than natural gas compressed to similar densities.[8]

For a decade, hydrogen-rich town gas was stored underground in an aquifer some 20 miles west of Paris, near the small town of Beynes. In the oil-shocked hydrogen community of the 1970s, Beynes carried a certain amount of name recognition for a while as a successful example of hydrogen storage. Ironically, this facility came to the attention of the hydrogen community only after French authorities were making plans to phase out the town gas and convert the facility to store natural gas in the middle and late 1960s. One problem with this storage method is that a fairly large amount of gas, be it natural gas or hydrogen, has to be "invested" initially and written off in order to build up the underground pressure so that the recoverable part can be squeezed out later. This "cushion gas" cannot be recovered easily. Reports at the time said that the Beynes field had a storage capacity of 7 billion cubic feet but needed an additional 6 billion cubic feet of cushion gas.

Cryogenic storage is the space-age approach to storing large amounts of hydrogen. As has already been noted, a Hamburg utility is testing storage of liquid hydrogen in conjunction with the operation of a stationary fuel cell. In recent years, cryogenic storage has been high on the agenda of the WE-NET program, Japan's ambitious multi-decade scheme to develop a "world energy network" with hydrogen as the energy currency of choice.

The white plumes that billow from the Space Shuttle's fuel tanks before a launch are mostly water vapor frozen by cold oxygen exiting from the orange external tanks and made visible in the heat and humidity of the ambient air. In addition, near the three main engines, small amounts of hydrogen are being circulated through the exhaust chamber assembly and then vented to keep the nozzle cold enough to prevent the metal from melting when the rocket engine is running. Cryogenic storage at below −423°F is efficient, and liquefied hydrogen is a high-energy fuel. So far, however, cryogenic storage by utilities seems remote.

A 1972 American Gas Association report titled "A Hydrogen Energy System" pointed out that, at normal pressure, liquid hydrogen takes up only ⅟₈₅₀ as much volume as gaseous hydrogen. (For liquid natural gas, the factor is more than 600.) Liquefying hydrogen is expensive, though. Commercial liquefaction processes consume about a third of hydrogen's energy content. There is promise for more efficient, less expensive refrigeration techniques with "adiabatic demagnetization," a line of research pursued also in connection with work on superconducting magnets. Liquid hydrogen has been used and stored in fairly large quantities (typically between 15,000 and 26,000 gallons) for industrial use, but not as a fuel. The storage tanks at the Kennedy Space Center are much bigger—850,000 gallons and 900,000 gallons, respectively, for liquid hydrogen and liquid oxygen. Because of liquid hydrogen's peculiar characteristics, those containers are expensive, costing between $2 and $4 per gallon of storage capacity, according to the 1975 IGT survey. (The numbers probably haven't changed much in the intervening years, one expert opined in 1999.) They are dewars—complicated double-walled structures, similar to a thermos bottle in concept but with a stainless steel or aluminum inner liner, a vacuum between the two shells, and a steel outer casing. They are spherical because with that shape the total surface is smallest in comparison to the volume of liquid stored, keeping evaporation losses to a minimum. Loss rates below 0.5 percent per day for big tanks have been achieved. With a vacuum inside, curved steel sections withstand atmospheric pressure better than straight walls. As with aquifer storage, some liquid hydrogen is always left in these tanks, not to build up pressure but to keep them permanently chilled. Once they are allowed to "heat up" to ambient temperature, it takes sizable quantities of expensive LH_2 to get the temperature down to storage levels again. The tanks are slightly pressurized to keep outside air from coming in; the air would freeze immediately, clogging up valves and other passages. Frozen solid oxygen from the air would present a hazard if it came into contact with hydrogen.

Aside from space applications, liquid hydrogen has not been widely used as a fuel. The double-walled piping and related equipment used in aerospace applications has been far too expensive for down-to-earth energy use. This may change if Japan's ambitious WE-NET goal of turning hydrogen into the world's principal energy currency picks up momentum. Part of the

WE-NET's agenda is the development of huge storage and transportation systems for liquid hydrogen. One component of that effort is the development of a 50,000-cubic-meter underground storage tank (almost 15 times the size of the Kennedy Space Center tank); another is the development of large liquid-hydrogen tankers. Also under development in the WE-NET program is a 500-megawatt liquid-hydrogen-burning gas turbine.

Hydride storage of hydrogen, in which hydrogen is stored inside the metallic lattice structure of certain alloys such as titanium-iron, has received more attention in automotive applications than in others, but in its early years it was a subject of intense interest for large-volume storage. Companies and institutions that investigated hydride storage for industrial- and utility-scale hydrogen in the 1970s included Allied Chemical, International Nickel, Phillips Research, Sandia National Laboratory, Battelle Memorial Institute, Brookhaven National Laboratory, and Public Service Electric & Gas of Newark, New Jersey. Today, the concept of large-scale hydrogen-storage via hydrides has been just about discarded as awkward and material-intensive; other storage methods, such as liquid hydrogen, seem more promising. But in the 1970s, quite a bit of research was carried out. Brookhaven National Laboratory, for example, developed fairly detailed engineering concepts for a 26-MWe peak-power electric plant driven by hydrogen stored in large iron-titanium hydride storage beds. The plant would have utilized an alkaline-electrolyte pressure electrolyzer, 10 iron-titanium storage beds, and a 24-module hydrogen-air fuel cell.

Aside from energy storage, hydride systems have been proposed and investigated for hydrogen purification, compression without normal compressors, separation of hydrogen isotopes to produce deuterium for heavy water, heating, air conditioning, refrigeration, heat storage, and recovery of waste heat. Hydride hydrogen compressors would do away with conventional compression equipment by simply heating a saturated hydride tank. With different hydride combinations and multi-stage operations, pressures of up to 100 atmospheres are possible, using only low-grade heat as energy source, according to an early paper by Frank Lynch and Ed Snape at the 1978 Zurich World Hydrogen Energy Conference.

Isotope separation, investigated at various times by Daimler-Benz, Brookhaven Lab, and General Electric, exploits the phenomenon that certain hydrides, such as titanium-nickel hydrides, preferentially absorb heavy hydrogen (deuterium). One Daimler-Benz study suggested large-scale separation processes in which deuterium would be extracted before normal hydrogen is sold as car fuel; deuterium's sale price could create a credit that would help bring down the price of hydrogen fuel.

Finally, the heat effects of hydrides—heat release during hydrogen absorption and heat takeup during hydrogen release, but at different levels and pressures for different alloys—open up prospects for heat storage, pumping, air conditioning, refrigeration, and power generation, Lynch and Snape reported. One of the first examples of such a system was the HYCSOS hydrogen conversion and storage system, which employed two hydride materials, lanthanum-nickel (LaNi) and calcium-nickel (CaNi), developed experimentally by Argonne National Laboratory in the mid 1970s. By selectively shuttling hydrogen back and forth among four tanks, exploiting different temperature and pressure gradients of the hydride materials plus solar heat input, the desired effects of cooling or heating a room were achieved.

Newer concepts for developing hydride-powered air conditioners for buses and for cars were developed in the decades that followed, but none of them made it to the commercial stage. In the late 1980s, for example, a team at the Carnegie Mellon Institute, led by W. Edward Wallace, designed a hydride automotive air conditioner. Wallace claimed General Motors and one Japanese car manufacturer were interested.[9] Two years later, two scientists at the Technion, the Israel Institute of Technology, were looking for funding to build and install a prototype hydride air conditioner in a bus.[10] Moshe Ron and Yale Josephy, who had been working on the idea since the early 1980s, said the device, to be powered by the heat from the exhaust gases, would reduce the fuel consumption of a bus by about a third, mainly because of the elimination of the power-hungry AC compressor.

In the mid 1990s, Thermal Electric Devices, a small company based in Albuquerque, tried to develop a hydride-operated refrigeration and air conditioning system in cooperation with the Westinghouse Savannah River Company of Aiken, North Carolina,[11] one of the most important US facilities for making nuclear weapons. As part of its weapons work, Savannah River had amassed a vast store of information on making and using hydrides. The process would be environmentally friendly, with no CFCs or

HCFCs used in the cooling-heating cycle. But in the end the project didn't pan out. "The essential problem was capital cost," one of the principals involved in the project wrote in a 1999 e-mail message. The concept may be revived again in the future, he added, but nothing concrete is on the horizon.

Cryo-adsorption, another mostly untried storage idea, is a compromise between cryogenic and hydride storage. Adsorption is a way of storing a gas or a liquid next to the surface of the adsorptive material, such as activated carbon or nickel silicate, rather than drawing it inside the storage medium (as in hydride storage). A cryogenic adsorption concept with temperatures around −320°F—the range used industrially in the liquefaction of nitrogen, and nowhere as low as those used in pure cryogenic storage—was presented at the 1976 Miami Beach hydrogen conference. Unlike cryogenic storage, which proceeds at ambient pressure, cryo-adsorption requires pressure of about 60 bars—low compared to the 1500–2000-bar pressures that have been suggested as necessary for efficient pipelining.

Compared to hydride storage, with its weight problems, cryo-adsorption looked good—on paper, at least. Proponents claimed that weight could be cut by as much as one-third compared to titanium-iron hydride, but volume requirements were higher by a factor of 3. In any event, nothing much has been heard about it in recent years, and it seems to have fallen by the wayside.

For most of the twentieth century, Germany's Ruhr region was synonymous with the country's industrial might—big steel plants, coal mines, big machinery, big guns. In the late 1980s, that luster took on a tired brown tinge. The once-mighty Ruhr, like old-line steel and coal industries in many countries, succumbed to the "rust belt" syndrome.

In that industrial territory along the banks of the Rhine are a number of large chemical plants producing basic chemicals that go into aspirin, food additives, dyes, fertilizers, plastics, and fibers. A dozen of these plants have the distinction of being linked to the world's oldest hydrogen pipeline network, and one of the most extensive: a 130-mile system of buried steel pipelines wrapped in bitumen and plastic that traverses cities, crosses the Rhine in two places, and transports more than 10.6 billion cubic feet of hydrogen per year (almost one-third of the total industrial hydrogen production of the entire grid).

The system had its beginning in the late 1930s with a humble 14.3-mile three-point system in which the newly formed chemical firm of Chemische Werke Huels began feeding hydrogen to two plants near Gelsenkirchen. Since then, the grid has expanded to link four producers of the gas and nine users. Remarkably, in more than 50 years of operation there has been no major accident from escaping hydrogen or potentially explosive hydrogen-air mixtures, a record that is even more admirable in view of the fact that the line crosses mining areas where ground settlements occur. Expansion joints placed into the line have successfully coped with stresses caused by soil shifts in subsiding, worked-out coal mines.

Suggestions to expand the pipeline network were made in the mid 1970s to take care of the expected increase in demand for hydrogen, especially for new hydrocrackers in the petrochemical industry, but nothing ever came of it. In 1993, the entire pipeline system was sold to Britain's BOC industrial gas company.

During the first burst of interest in hydrogen energy that followed the Arab Oil Shocks, the grid enjoyed something approaching celebrity status among advocates of hydrogen energy. They pointed to the Huels pipeline as proof that hydrogen can be transported safely and economically like natural gas, delivering clean energy over large distances at a fraction of the cost of delivering energy via electricity and with less "visual pollution."

A 1993 study by Ludwig-Boelkow-Systemtechnik that looked at international hydrogen and hydrogen pipelines from the perspective of steelmaking for the Euro-Quebec Hydro-Hydrogen Pilot-Project (EQHHPP) found that 469 miles of hydrogen pipelines were in operation in half a dozen countries. Even bigger than the German system was Air Liquide's system in France, with 212 miles. In the United States, Air Products and Chemicals has been operating a 62-mile pipeline in the Houston area since 1969. Pipeline diameters ranged from 30 mm (Los Alamos, New Mexico) to 273 mm (Alberta, Canada), and pressures from 3400 kPa (Air Products, Louisiana) to more than 100,000 kPa (Rockwell International, near Los Angeles).

In the 1960s and the 1970s, when nuclear power was still untarnished and in full bloom, Cesare Marchetti and other energy strategists argued that massive nuclear plants, which were then thought likely, would have to be located far from population centers. But beyond 1000 miles or so, transporting energy via electricity would become disproportionately costly, they said; the costs of transporting energy via hydrogen would increase less drastically.

Though the basic feasibility and the attractive economics of hydrogen pipelines are not in dispute, the details are in dispute because of the myriad of variables that come into play, including pipeline diameters and pressures, spacing of compression stations, materials, embrittlement, fuel costs for pumping stations, types of compressors available, and the geographical locations of sources of hydrogen.

Embrittlement is still an ongoing concern for both gaseous and liquid hydrogen. Steel becomes brittle when it is exposed to hydrogen; the higher the pressures and the higher the temperatures, the more pronounced the problem. The EQHHPP study noted that hydrogen generally "can have a detrimental effect on toughness, ductility, burst strength and fatigue life," and that "pipeline steel is exposed to embrittlement already at normal temperatures." At the 1996 World Hydrogen Energy Conference, a number of reports dealt with materials and safety, and German and Japanese scientists presented papers that looked at hydrogen embrittlement of austenitic stainless steels. The German study concluded that higher nickel content made such a steel more susceptible to hydrogen embrittlement. The Japanese study found that the ductility of certain weld metals decreased remarkably at cryogenic pressures, but that other steels were still sufficiently strong in their mechanical properties to be used in these harsh environments.

One method of preventing embrittlement cracks in future hydrogen pipelines may be the addition of very small amounts of other chemicals, such as oxygen or carbon monoxide; more studies are apparently needed. It has also been suggested that other additives should be mixed in as a safety measure that would either give a characteristic smell to normally odorless hydrogen or add a distinctive color to the normally invisible hydrogen flame—a warning to the user that hydrogen is around. This would be fine as long as hydrogen was used only as an energy carrier; impurities generally don't interfere with the burning process. But many chemical processes are dependent on extremely pure hydrogen. Chemical manufacturers presumably can pay more for hydrogen used as an input in the production of higher-priced chemical products, it has been argued, thereby creating a bigger market that in turn would help to lower the costs and thereby help promote the more rapid phasing in of hydrogen as a fuel. But if "denatured" hydrogen is useless to the industry, that particular synergy is lost.

In the 1970s, the Texas Gas Transmission Corporation funded research into hydrogen production from water because, it said at the time, it wanted to preserve its huge investment in natural gas pipelines by converting them to transport methane made with hydrogen. One early estimate of the total value of all existing pipelines in the United States put the total at about $40 billion in those years. However, a 1976 study by the Stanford Research Institute found problems with that estimate: the geography was all wrong. It noted that the trunk natural gas pipeline network fanned out from the major gas-producing areas, which were concentrated in the Gulf Coast region, terminating or tapering off in capacity with increasing distance from the gas fields. However, the study noted, there was no natural reason to concentrate major hydrogen-producing facilities in the Gulf Coast region. It was unlikely, therefore, that conversion to hydrogen would protect those investments, since the then-extant pipeline network generally did not link the correct places and since it extended in inappropriate directions.

In the late 1980s, an indication of the size of the investments required for long-distance transport of energy—hydrogen, electricity, or both—was provided by a detailed theoretical analysis of shipping solar energy from North Africa to West Germany.[12] The study assumed conversion and transport of high-voltage direct current (800 kV on land, 500 kV underwater) by cables traversing the Straits of Messina, pipelining of gaseous hydrogen, and a combination of the two. The idea was to assess the cost of shipping this energy from Algeria to Nordrhein-Westfalia, home state to the industrial Ruhr. The study assumed that the solar farm was somewhere in the middle of Algeria, some 1500 km from the Straits of Messina and another 2500 km from the German terminal. Another assumption was that the delivery system would be optimized to match Nordrhein-Westfalia's particular demand pattern, in which 53 percent of the total was met by coal. A third assumption was that the equivalent of 34 million tons of coal—two-thirds of the region's energy consumption, met at the time by imported oil and natural gas—would be replaced by solar energy from

Algeria. The conclusion was that a combination of piped hydrogen and transmitted high-voltage electricity provided the lowest transport cost because it offered the smallest transport and conversion losses: 18 percent, versus 24 percent for pure electric transmission and 27 percent for piped hydrogen. Pure high-voltage transmission came in second, mostly because the authors estimated the electricity portion of the total energy required for the region at about 20 percent. If the electric portion were smaller, hydrogen transport would have been judged cheaper. Either way, the calculated investment costs for such a system would be enormous and probably prohibitive: the total for the combination system came to DM 245 billion (at 1988 prices—$144 billion at the autumn 1997 exchange rate of DM 1.70 per dollar). A "pure" hydrogen pipeline system to carry the entire load would have required investments of DM 275 billion ($161.8 billion). The high-voltage-only system would have been the most expensive—a staggering DM 343 billion ($201.8 billion).

It is possible that massive long-distance pipeline systems for hydrogen will not be needed—at least not for the initial decades in a future shift to hydrogen energy, and not in the United States. Certainly, the construction of new hydrogen pipelines for utilities and power plants looks highly unlikely for the foreseeable future. The question has been raised in the context of fueling cars and buses and setting up a supply infrastructure for zero-emission vehicles cars in California, where zero-emission legislation will soon take effect.

In the 1990s, one widely recognized expert concluded that hydrogen pipelines would not be needed to meet the fuel needs of a first generation of fuel-cell-powered zero-emission light-duty vehicles in the first few years of the twenty-first century. In a 1993 study on renewable hydrogen energy systems, Joan Ogden of Princeton University's Center for Energy and Environmental Studies observed that there were "good to excellent" resources available for renewable hydrogen production "globally and in most areas of the United States."[13] Ogden, who studies the economics and system requirements of solar hydrogen, found that with photovoltaics-driven electrolysis alone it would be possible to fuel all light-duty vehicles—cars, light trucks, and presumably even gas-guzzling sport-utility vehicles—operating in the United States circa 2010, if all the vehicles would be powered by high-efficiency fuel cells. The production of this hydrogen would require about 0.1 percent of the land area of the United States, equal to 1 percent of the desert areas. If wind power were to be used as a hydrogen source, Ogden calculated, the percentage would increase to 2 percent of the total land area. With even less efficient biomass as a hydrogen source, the land area required would increase to 3 percent of the country (equal to 70 percent of the cropland currently idle, Ogden noted). "Because good local renewable hydrogen resources are available in most parts of the US, it might not be necessary to build long distance hydrogen pipelines," Ogden wrote. Elsewhere, the picture might be different, though: "It might be necessary in areas such as Northern Europe to transmit hydrogen long distances." For a large 1000-mile pipeline transmitting the equivalent of 0.5 gigajoule of energy per year, Ogden calculated, the delivered cost of that hydrogen would be about the same as for a local city supply—about $19.6–$27.3 per gigajoule, or about $2.55–3.56/gallon gasoline equivalent—because increased transmission and distribution costs would be offset by savings in large-scale storage and compression costs. But since fuel cell vehicles are several times as efficient as i.c.-engine vehicles, the higher per-gallon/gasoline-equivalent price for hydrogen wouldn't matter much. In Ogden's calculations, the break-even gasoline price, including tax, would be between $1.29 and $1.60—which is what American drivers were paying during much of the 1990s. In general, Ogden wrote, "even though renewable hydrogen would be several times as expensive as gasoline, and hydrogen fuel cell vehicles would probably cost considerably more than gasoline vehicles, our analysis suggests hydrogen fuel cell vehicles might compete on a lifecycle cost basis," because, in addition to higher efficiency, they would have longer lifetimes and lower maintenance costs.

Based on her extensive review of the literature and on experience with hydrogen pipelines, Ogden is convinced that hydrogen can be safely piped. In regard to embrittlement Ogden said the following in a late-1997 interview: "You can't rule out embrittlement if you put through pure hydrogen at 1000 psi. Welds and other components are susceptible to it. The problem is really crack propagation under pressure cycling—making the pipes flex propagates the cracks." The addition of small amounts of trace gases—oxygen, carbon dioxide, or carbon monoxide—seems to inhibit that. "If you build the pipeline [with hydrogen in mind] from the start, you can probably build it right," she added.

According to Ogden, the scale of energy flows in today's systems is still very much lower than that of natural gas, "but enough hydrogen is shipped around already in pipelines to fuel several million fuel cell vehicles."

10

Non-Energy Uses of Hydrogen: Metallic H₂, Biodegradable Plastics, and H₂ Tofu

"To the no-nonsense engineer, any talk of a 'hydrogen economy' may seem like so much hot air," said the pithy lead sentence of a May 1992 story in *Chemical Engineering* that detailed the growth of hydrogen use because of heightened environmental concerns. The opening paragraph, headlined "'Green' Laws Spark Hydrogen Technologies," quoted an Air Products and Chemicals manager, Michael Hilton, as saying that "as legislative, safety and environmental issues continue to tighten, they're promoting hydrogen's chances as an energy source and, more immediately, its prospects as a chemical feedstock." In the coming decades, the article explained, increasingly tough environmental regulations are likely to translate into a greater need for hydrogen as an alternative to fossil fuels just when traditional sources of commodity hydrogen are drying up, with most petroleum refiners' "hydrogen balances" (i.e., hydrogen produced as by-product) going negative. "Previously, gasoline was made with benzene which means that H_2 was rejected," the magazine said. "But now that the laws mandate lower aromatic and higher oxygenate levels in gasoline, there's less H_2 available as by-product. At the same time, H_2 demand is rising in hydrodesulfurization units, since the same laws require refiners to cut sulfur levels in fuels." Worse yet, the magazine reported, "supplementary sources for the gas are also shrinking." A case in point is by-product hydrogen from the chloralkali industry; output is dropping "as demand for its coproduct chlorine weakens." And, in yet another twist, "H_2 demand for the making of hydrogen peroxide is growing as that environmentally safer bleach gains chlorine's market share."

Hydrogen is a very good chemical reducing agent. In chemistry, reduction is essentially the reverse of oxidation, the process in which materials react with the air's oxygen to form another compound. Rust (iron oxide) is the product of the interaction of iron and atmospheric oxygen; it can be transformed into iron again by exposure to hydrogen. (In industrial iron production, ores are reduced to metallic iron in a smelter.)

In the international chemical industry, hydrogen has been for decades one of the most important chemical raw materials for the production of many organic materials. Hydrogen plays a role in making ammonia-based fertilizers, in refining crude oil into petroleum, and in making methanol (which, in addition to its recently discovered potential as a reasonably convenient hydrogen carrier for fuel cell cars, is the basis for resins, varnishes, plastics, solvents, and antifreezes).

In the oil industry, refineries and other plants routinely produce and consume hydrogen. Hydrogen is needed in "hydrotreating" processes, in which sulfur and other impurities are removed during distillation of crude oil— and its importance is growing. An overview paper on the industrial uses of hydrogen presented at the 1996 World Hydrogen Energy Conference noted that "due to the increased use of heavier crude oils, containing higher amounts of sulfur and nitrogen and to meet stringent emission standards, need for hydrogen is experiencing a very rapid growth in the petroleum refining industry." The authors of this paper, Ram Ramachandran, Raghu Menon, Raymond Morton, and Thomas Bailey, of BOC Gases, said that industrial use of hydrogen can be broadly divided into the following four categories:

As a reactant in hydrogenation processes—here hydrogen is used to produce lower molecular weight compounds or to saturate compounds or to crack hydrocarbons or to remove sulfur and nitrogen compounds.
As an O_2 scavenger—to chemically remove trace amounts of O_2 to prevent oxidation and corrosion.
As a fuel in rocket engines and potentially in automobiles.
As a coolant in electrical generators to take advantage of its unique physical properties.

Furthermore, in "catalytic cracking" of crude oil, hydrogen is produced as a by-product. In "hydrocracking," large oil molecules are broken down with hydrogen into fuel distillates that are blended into gasoline.

There are many industrial uses for hydrogen. Hydrogenation of edible organic oils made from soybeans, fish, cotton seed, peanuts, corn, and coconuts slows their propensity to oxidize and turn rancid. The addition of hydrogen converts a liquid oil into a solid fat, such as margarine or shortening. Inedible tallow and grease treated with hydrogen can be used to produce soap and animal feed. According to Ramachandran et al., these processes are typically carried out in the presence of nickel catalysts.

In the manufacture of polypropylene (a type of plastic), hydrogen is used to control the molecular weight of the polymer, according to Ramachandran et al. In a more recent application, hydrogen is used in plastics recycling: the plastic materials are melted, then the molten plastic is hydrogenated to crack it to produce lighter molecules which can again be reused to produce polymers. Ramachandran et al. note that "as the environmental regulations and consciousness of the public grow, this may become more popular."

One of the most widespread uses of hydrogen is to make ammonia fertilizer by means of a high-pressure reaction between nitrogen and hydrogen. Ammonia production consumes about 59 percent of the hydrogen produced in the world at present, according to Ramachandran et al.

In metallurgy, according to Ramachandran et al., hydrogen is used in the reduction stage in the production of nickel in a process known as the Sheritt Gordon Process. In electronics, hydrogen is used in the "epitaxial" growth of polysilicon by wafer and circuit manufacturers.

Hydrogen, both in pure form and mixed with nitrogen, is used as an oxygen scavenger in metalworking. Hydrogen atmospheres are used in the heat treating of ferrous metals to change some of their characteristics—for example, to improve ductility and machining quality, to relieve stress, to harden, to increase tensile strength, to change magnetic or electrical characteristics. A hydrogen-nitrogen atmosphere is used in annealing steel to make it more machine-workable, to make it amenable to cold rolling, and to reduce stress while the metal is being shaped or welded. Bright annealing makes for a smooth, shiny surface, like stainless steel. A hydrogen-nitrogen atmosphere is used to reduce or prevent oxidation and annealing of nonferrous metals as well. A reducing atmosphere of relatively pure hydrogen is used in processing tungsten, in producing molybdenum, and in producing magnesium via electrolysis from magnesium chloride.

Burning hydrogen with oxygen produces very high temperatures for the cutting of glass and for the cutting and high-temperature melting of quartz, and in "float glass manufacture" (where the glass typically floats on a tin bath) a mixture of 4 percent hydrogen in nitrogen is used to prevent oxidation of the molten tin.

In the manufacture of electronics components such as vacuum tubes or light bulbs, brazing (a process in which materials are heat-bonded) is carried out either in a hydrogen atmosphere or in nonreactive gases (such as argon or nitrogen) to prevent oxidization.

Hydrogen's unique physical properties also have applications in the electric power industry. Hydrogen has the lowest viscosity among fluids, Ramachandran et al. report, and therefore it is best suited to reduce friction in rotating power equipment. Gaseous hydrogen is used to cool large generators, motors, and frequency changers. Hydrogen has a greater thermal conductivity than normal air and therefore provides better cooling. A hydrogen cooling system is a closed circuit in which the gas is routed via heat exchangers through the generator shell and through the stator windings. (There is some leakage.)

In nuclear research, liquid hydrogen is used to fill bubble chambers to make the traces of subatomic particles visible and photographable. In the nuclear power industry, hydrogen is used in some stages of nuclear fuel processing.

Hydrogen's ability to reduce ores to their metallic states as an alternative to conventional smelter methods has attracted the attention of researchers for more than two decades. One of the earliest proponents of this method was Tokiaki Tanaka of Hokkaido University, a specialist in non-ferrous metals, who presented a detailed review of hydrogen's potential in ore reduction in the December 1975 issue of the *Journal of Metals*. Taking his cue from work on thermochemical hydrogen production cycles under way at the time at the Euratom research center in Ispra, Italy, and elsewhere, Tanaka said that the chemical reactions in these cycles dovetail nicely with the processing of sulfur-containing sulfide ores. As Tanaka saw it, the development of these non-polluting methods of metal production were environmentally more important than the production of hydrogen per se. Tanaka saw a particularly great potential benefit in the use of hydrogen in copper smelting, which now takes three or four energy-intensive steps. Tanaka concluded that "future developments in the field of hydrogen technology may . . . bring about a revolution in the field of metal extraction as we now know it."

Hydrogen's potential uses in steel-making and (strangely symbiotically) in helping to preserve Brazil's rain forests were pointed out in the early 1990s by Lutero Carmo de Lima, a researcher with the Department of Mechanical Engineering at Brazil's Federal University of Uberlandia. De Lima suggested that electrolytic hydrogen could replace much of the charcoal made from rainforest wood then being used to produce iron and steel in Brazil, where charcoal-based iron and steel production then accounted for about 26 million tons out of a total annual output of about 40 million tons. According to de Lima, some 20 percent of Brazil's annual energy consumption came from the burning of wood. (The biggest shares of Brazil's energy consumption came from hydropower and imported oil, at about 30 percent each.) De Lima, who at the time was working at the University of Miami's Clean Energy Research Institute, asserted in a paper that construction of several large hydropower dams in Amazonia could provide low-cost electricity, which eventually could be tapped for hydrogen production. Realistically, however, he did not expect hydrogen to be cost-competitive with fossil fuels until about 2010; large-scale substitution of hydrogen for charcoal, he noted, probably wouldn't begin until about 2020.

In the early 1990s, the Euro-Quebec Hydro-Hydrogen Pilot Project (EQHHPP) embarked on a project to launch hydrogen-powered steelmaking with a small pilot plant in Ireland. The project's key movers were Joachim Gretz (then an official at the Joint Research Center of the European Community in Ispra), Willy Korf (a German steel executive with a reputation in the industry as a maverick), Tom Doyle (a former official of the European Community), and Raymond Lyons (the director of Industrial Consultants International and a director of the Irish firm Kent Steel). Korf apparently had planned to introduce steel reduction via hydrogen technology in his plants, but he died in a plane crash before he had a chance to carry out that plan.

In a 1990 article in the German solar energy publication *Das Solarzeit-alter*, Gretz, Korf, and Lyons noted that in conventional steelmaking each kilogram of molten steel releases about 2.2 kilograms of carbon dioxide into the atmosphere, and that the international steel industry contributes about 10 percent "of all carbon dioxide emissions caused by man with the combustion of fossil fuels." "Hydrogen," they suggested, "is an excellent and clean means of reduction that produces steam rather than CO_2, and it does not generate any additional impurities as does coke (especially sulfur)."

The overall goal of EQHHPP's project was to set up hydrogen-fueled steel plants in some portions of the former Soviet Union (including Siberia), in Brazil (with its huge hydroelectric potential), and in Ireland. In late 1991, Kent Steel began making plans for a small demonstration plant in Cork, with management assistance from Ludwig-Boelkow-Systemtechnik (a nonprofit consulting group that for years had managed various parts of EQHHPP) and from Planit Associates. Kent Steel also proposed a second, bigger facility in Arigna, Ireland, that would process 3 tons of iron ore per hour. Neither facility ever made it beyond the brainstorming stage.

Elsewhere, in a different technological context, the use of hydrogen-rich gas in the direct reduction of fine iron ores is already an industrial reality. According to an article in the April 7, 1997 issue of the German industry publication *Stahl und Eisen*, several plants using hydrogen as a reducing agent were already in operation, and a two-stage direct reduction–electric arc method of steel production had good prospects. The authors, Heinrich Wilhelm Gudenau (head of the Institute for Smelter Technology at Aachen Technical University) and Martin Hirsch (general manager for process development at Lurgi Metallurgie in Frankfurt), reported that in the aforementioned two-stage method iron ore is reduced directly to "sponge iron" with the help a hydrogen-rich reduction gas produced from natural gas. The ore-melting phase is bypassed entirely. In the second step, sponge iron plus scrap metal is melted in an electric arc. Two such processes, the so-called Midrex and Hyl III processes, were technically mature and were already producing about a million tons of sponge iron a year. Several production facilities were already in operation, under construction in Venezuela and Australia, or planned, with capacities between 400,000 and 2 million tons a year. An advanced variant, a "fluidized bed" process dubbed the Fine Iron Ore Reduction (FIOR) method, was at the time of the article's publication the only industrially operating process of its type, producing 400,000 tons of "hot briquetted" iron. Developed by Exxon, it was in use in Venezuela; more advanced versions were said to be on the drawing boards.

Atomic, solid, and metallic hydrogen, researched since the late 1960s in the United States, in the Soviet Union, in Europe, and in Japan, have a "laboratory curiosity" status today similar to that of liquid hydrogen in the 1930s, the 1940s, and the 1950s. Researchers have speculated about the properties of these so far hypothetical materials. (Hydrogen in metallic form is believed to form the core of the planet Jupiter.) Of course no industrially useful quantities have been produced, but if they could be produced they might be useful as ultra-powerful rocket fuels or as room-temperature superconductors.

Atomic hydrogen (H) exists only in infinitesimally small percentages in normal gaseous hydrogen, which is almost always molecular (H_2) because the bonds between hydrogen atoms are very strong—stronger than any other chemical bond. If the two hydrogen atoms are somehow separated, their natural tendency is to immediately recombine into the H_2 molecule. This strong attraction and the energy implied by this bond have persuaded the US Air Force to try to break up the hydrogen molecule as a means of storing energy. When recombining into molecular hydrogen, atomic hydrogen would yield about four times as much energy as the best currently available chemical rocket fuel (liquid hydrogen combusted with liquid oxygen).

Since the early 1970s, NASA's Lewis Research Center in Cleveland has been working on methods of turning a percentage of gaseous molecular hydrogen into atomic hydrogen. Initially, this effort was based on the research of a German scientist, Rudiger Hess. Hess's 1971 doctoral dissertation detailed his efforts, in which he split the hydrogen molecule into two atoms by running the gas through a pair of electrodes. According to a 1974 article in *Aviation Week and Space Technology*, Lewis researchers cooled a mixture of molecular and atomic hydrogen down to 4° Kelvin (barely above absolute zero), and the mixture then condensed out on the walls of the experimental apparatus. The entire setup functioned inside a superconducting magnet field that, theoretically at least, aligned the electrons in the hydrogen atoms in such a way that they could not recombine into molecules. The researchers believed that eventually a method might evolve by which the atomic-molecular hydrogen mixture could be stable, though very low temperatures would be necessary.

Walter Peschka, the scientist who had converted BMW sedans to run on liquid hydrogen, was among the first to investigate these phenomena. (Rudiger Hess was one of his students.) In December 1978, Peschka managed to store 2 grams of spin-aligned atomic hydrogen for several hours. Peschka estimated the energy that could be stored and released by reverting to the molecular state to be 10–20 times the amount available from current chemical fuels.

Another exotic variant is solid hydrogen. First predicted by Eugene Wigner and Hillard Huntington in the early 1940s, solid hydrogen was actually produced in microscopic quantities in 1989 by two scientists at the Carnegie Institution's Geophysical Laboratory, Ho-Kwang (David) Mao and Russell Hemley, by squeezing samples of gaseous hydrogen under ambient conditions. (Ten years earlier, Mao and Peter Bell had produced solid crystalline hydrogen from liquid hydrogen at room temperatures and at a pressure of 650,000 bars.) In the June 23, 1989 issue of *Science*, Mao and Hemley reported that they had subjected hydrogen to enormous pressures—more than 2.5 megabars (2.5 million atmospheres). On the basis of optical measurements of light transmission and reflectivity, they claimed to have produced a "semi-metal"—a substance that conducts electricity, but not as well as a "real" metal.

Metallic hydrogen is a variant of solid hydrogen created by pressures even higher than those required for "mere" solid, crystalline hydrogen. When molecular solid hydrogen is subjected to pressure, the hydrogen molecules, normally spaced about 10 times as far from each other as the two hydrogen atoms forming the molecule, move closer together. Eventually, the molecules are so close together that the distance is about equal to the atoms' distance from each other inside each molecule. At that point, the atoms of different molecules begin to interact. If the pressure goes still higher, a metal eventually forms.

Solid "metallic" hydrogen occurs under extremely high pressures—maybe a million atmospheres or more. Apparently it isn't certain at precisely what pressure the change from mere solid to metallic hydrogen occurs. It may be that not only hydrogen turns metallic. Some researchers theorize that any material—"beach sand, a plastic spoon, a mug of beer, even the air you breathe" would "become a shiny metallic solid like aluminum or copper at very high pressure," Robert Hazen, a research scientist at the Carnegie Institution and a professor of Earth Sciences at George Mason University, has written.[1]

Theory postulates that metallic hydrogen may have the same energy content as spin-aligned atomic hydrogen—about 50,000 calories per gram (1,400,000 calories per ounce)—but that, because it is assumed to be 14 to 15 times denser, it would pack a bigger wallop in a smaller space.

Mao's and Hemley's account created some controversy because of the apparently erroneous perception that they claimed to have produced metallic hydrogen. In his book *The Alchemists*, Hazen notes that this debate raged for years among experts on high-pressure physics. He asks: "Does the observed darkening above 2.5 megabars correspond to the metallization long predicted by scientists, or is it the sign of some other exotic material? When metallization is confirmed, will scientists be able to stabilize the metal at room temperature?" "What does seem certain," Hazen continues, "is that hydrogen, compressed to unimaginable pressures between the flat faces of diamonds, forms a substance that is quite unlike anything anyone has seen before." Metallic hydrogen could be "a superconductor at room temperatures," Hazen wrote, echoing theories that had been circulating for decades. It could be the "ultimate electronic material." Theorists calculate that it also could be the most concentrated form of chemical energy. "As a rocket fuel," Hazen writes, "it could store hundreds of times more thrust per pound than any other material; as an explosive it would be thirty-five times more destructive than TNT."

Apparently, the US military establishment has pursued—and may still be pursuing—the production of metallic hydrogen. "There could be no better way to pack hydrogen atoms into a hydrogen bomb than in its dense metal form," Hazen notes. "Keenly aware of such vast potential for destruction, some scientists raced to produce metallic hydrogen with a new dark urgency. They were driven not so much by curiosity about the unknown as by fear that others might exploit the substance first." But many unanswered questions remain: "Once made, will the metal persist after pressure is released? Could you hold a chunk of metallic hydrogen in your hand, or would it be unstable and revert immediately to the gas form? Only the scientists in high-pressure laboratories can provide the answers."

In a 1989 telephone interview, Mao told me that the solid hydrogen he and Hemley had produced had returned to the gaseous state after the pressure was removed. Other theorists had speculated the material would remain stable or at least metastable. Mao's former collaborator Peter Bell had speculated that metallic hydrogen could be made stable by "hydriding" it—adding small amounts of another metal, such as palladium, to turn it into an alloy. Mao said in the interview that this is a possibility, depending on whether the material is "quenchable"—the technical term for arresting the material in its altered state.

A radically different approach to making metal hydrogen has been pursued for almost 20 years by scientists at Lawrence Livermore National Laboratory in California: firing projectiles from huge guns at liquid-hydrogen-filled containers to very briefly subject the hydrogen to extreme pressures. As was first reported in 1996,[2] William Nellis and his colleagues used a 60-foot-long two-stage gun. According to the *New York Times*, "for a fraction of a second . . . the impact converted the molecular liquid hydrogen" into an "excellent conductor (of electricity), presumably a metal."

Fittingly, hydrogen gas was used in the gun's propellant system. In the first 30-foot stage, ignited hot gunpowder gases shot a piston forward, which in turn drove hydrogen gas ahead of it. The hydrogen gas reached a very high pressure before bursting a disk-type rupture valve at the end of that first tube. That high-pressure gas then shot into a much thinner second tube holding the plastic-and-metal projectile, which then slammed into the liquid-hydrogen-bearing target, having reached speeds of up to 9 miles per second (more than 32,000 miles per hour), generating a pressure of 1.4 million megabars. The drop of resistivity in the LH_2 sample had to be measured in about a ten-millionth of a second before the pressure dropped again.

One important aspect of the findings—that conversion to a metallic state occurred at pressures much lower than predicted—may help explain why Jupiter has a huge magnetic field. The *Times* quoted Nellis as saying it meant that the hydrogen in the planet's core becomes a metal much closer to the surface than previously believed, and that the upper 10 percent of the planet is much more electrically conductive than astronomers and scientists had thought. According to Nellis, "metallic liquid hydrogen at a relatively shallow depth in Jupiter would help account for Jupiter's gigantic magnetic field, some 20 times greater than the earth's."

These efforts to reach extremely high pressures began decades earlier at Cornell University, at the University of Maryland, at the University of Osaka, and at the Institute of High Pressure Physics in Moscow. A brief 1975 account by the Moscow team, headed by L. F. Vereshchagin, said the Russians had created what they thought was metallic hydrogen at temperatures of 4.2°K and pressures of about a million atmospheres. Their indication that metallic hydrogen had been created was also a change in electric resistivity: Hydrogen normally is a dielectric—an electric insulator—but the pressurized material "jumped" by some six orders of magnitude in terms of resistivity, an indication that the material acted like a true metal. "We note in conclusion that the return of the hydrogen to the dielectric state after the removal of the pressure cannot be regarded as a demonstration that metallic hydrogen cannot exist at normal pressures," Vereshchagin wrote, employing a strange double negative.

The Soviet work and the Japanese efforts, published almost simultaneously, were met with skepticism. Robert Hazen called both claims "spurious." It was suspected, he wrote, "that the drop in resistance ascribed to metallization was simply a short in an electrical circuit that had been crushed and drastically deformed by pressure." Atomic hydrogen, on the other hand, can be created without pressure by other esoteric techniques, such as glow discharge in conjunction with a strong magnetic field at extremely low temperatures (a method used by Peschka) or tritium decay ("the method of choice for most of the most recent experiments" at Lawrence Livermore National Laboratory, according to a 1993 overview paper by Bryan Palaszewski of the Lewis Research Center). Other technologies employ electron or other high-energy beams, nanotechnology, or microlasers.

Atomic hydrogen is not a solid but a gas. However, because it is spin-aligned, atomic hydrogen does not recombine into molecules, owing to the opposing electrical forces. This hydrogen state also displays characteristics, such as certain magnetic effects, that are expected to occur in metallic hydrogen. Experiments in making, storing, and handling tiny amounts of atomic hydrogen, some of them funded by NASA and the Air Force, have been going on for decades. Laboratories in the former USSR are also experimenting with this material.

Bryan Palaszewski's summary described the fiendishly difficult challenges of making these otherworldly materials in usable quantities, but also some of the promises of atomic hydrogen: "Modern experiments use nanogram samples of atomic hydrogen, whereas up to many hundred tons may be required for each launch from Earth to orbit," he wrote.

The energy stored in atomic hydrogen would be much higher than what is now available from conventional rocket fuels. Theoretically, the specific impulse[3] generated by the recombination of atomic hydrogen into its molecular form, without any chemical combustion via oxygen, could range from 600 to 1500 seconds of specific impulse. (The specific impulse of the Space Shuttle, powered by liquid hydrogen and liquid oxygen, is 453 seconds.) Put another way, a monatomic hydrogen-powered rocket could carry from 14 to 600 percent more payload, or the rocket's gross liftoff weight could be reduced by 82.7 percent, according to Palaszewski's paper.

Because of the expected difficulties associated with storing atomic hydrogen (for example, it likely will require huge magnets operating at cryogenic temperatures to prevent the spontaneous recombination of the hydrogen atoms), atomic hydrogen power may be best suited for Earth-to-orbit flights, like the Shuttle today, but with the magnets staying on the ground after liftoff. Flights requiring an upper stage may not be feasible.

One of the least-publicized and least-explored potential non-energy applications of hydrogen is in the fight against cancer. A pioneer in this field was William Fife, a professor of hyperbaric medicine at Texas A&M University. Fife, who had been investigating hyperbaric hydrogen (hydrogen under pressure as "hydrox," a breathing mixture of hydrogen in a safe, non-explosive ratio such as 95 percent hydrogen and 5 percent oxygen) for more than 20 years, said in a mid-1990s summary overview that the study of the effects of hydrox on living organisms was still in its infancy, with no more than 25 scientists active in France, Sweden, Canada, and the United States. "It must be emphasized that the field of hyperbaric hydrox for medical application is such a virgin field that virtually no definitive work has been done," he continued. "Thus, although our laboratory has carried out several preliminary studies, few scientists, and no industrial organization, have realized the potential for medical applications of this gas mixture."

Aside from medical applications, there has been little interest in the use of hydrogen-oxygen gas mixtures and no other research in the United States, largely because of the perception of danger, Fife added. An exception is the diving industry, which is considering hydrogen-oxygen gas mixtures for divers. Here, the most active, best-financed program is apparently one in Marseille, where the Comex company is conducting extensive tests, some in open water, for possible commercial operations.[4]

In a rare instance of media attention to Fife's work, the September 22, 1975 issue of *Medical World News* carried this one-paragraph item, headlined "Hydrogen Kills Tumors," in its "Outlook" section:

An experimental treatment using hyperbaric hydrogen has caused regression of squamous (medical synonym for "scaly") cell carcinomas in mice. Baylor College investigators Malcolm Dole and F. Ray Wilson and William P. Fife of Texas A&M University put three mice into a hyperbaric chamber (along with food and water) which was then flushed with a mixture of 97.5 hydrogen and 2.5 percent oxygen at a pressure of 8.28 atmospheres. After an initial ten-day exposure, the tumors turned black, some dropped off, some appeared shrunk at their base and to be in the process of being "pinched off." None of the tumor effects were noted in control mice kept at normal room temperatures. Continual remission of multiple carcinomas were shown in mice returned to the chamber for another six days, Further research will study the permanency of these results.

Dole, Wilson, and Fife, who subsequently published their basic findings in the October 10, 1975 issue of *Science*, were not quite sure what had happened. In a March 1976 letter, Fife said "this work is very preliminary and we do not yet understand the mechanism, or, indeed, the full scope of its potential." One possibility, Fife reasoned, was that molecular hydrogen under pressure scavenges "free radicals which are well known to cause some forms of cancer." Free radicals are atoms or multiatom molecules that "possess at least one unpaired electron": the atom or molecule is not internally balanced in terms of its electron-proton ratio but has excess electrical binding energy. "One hypothesis," said Fife, "is that destruction of these free radicals makes it possible for natural but weak immune systems of the body to cause the malignancy to regress."

In 1978, Fife said the earlier work may have been "somewhat inconclusive." He suggested that what his team saw disappearing may actually have been "papilloma (a precancerous wart-like cell condition) and maybe not the carcinoma." Nevertheless, work was continuing, though at a fairly slow rate. Taking another tack, Fife and his colleagues followed up their earlier work with research on the use of hyperbaric hydrogen in conjunction with other chemotherapy agents used in cancer research, such as Negromycin (then a new, very powerful, and very rare agent). Fife obtained about 4 grams of the material in 1978 from the National Cancer Institute; he was told this represented about half the world's supply then. He explained:

There seems to be some kind of synergistic effect in that the use of both of these together does extend the life of rats which have leukemia. . . . Normally, Negromycin causes the animal to die. It may cure the malignancy but it causes them to die with some of the symptoms similar to radiation damage, such as destruction of the bone marrow, over a period of 30 days. The few animals which were looked at histologically and that had the hyperbaric hydrogen did not show the bone marrow destruction, although the animals died from other causes. This has just not been worked out.

As to the perception of danger when mixing hydrogen and oxygen, Fife "categorically" stated in his mid 1990s summary paper that hydrox is safe to use. The Texas A&M Hyperbaric Laboratory had accumulated more than 6000 hours of exposing animals and humans to hydrox, at pressures ranging from 7 atmospheres absolute (ata) to 31 ata, for durations of up to 125 hours of continuous exposure, "without a single serious problem and without any accident related to the use of hydrox." Equipment needed "to mix, handle and use this gas mixture is either off-the-shelf or can be made without great difficulty." Research and development on commercial equipment for hydrox use needed "some attention, but there appear to be no particularly difficult problems which must be solved." In a 1970s experiment, Fife himself had lived on a non-explosive hydrogen-oxygen mixture at a simulated depth of 200 feet for 3 hours and at a simulated depth of 300 feet for 2 hours. "I cannot tell the difference between hydrogen and helium when breathing it, and I could notice no ill effects," he wrote.

In general, Fife recommended that the preliminary experiments of the early years be repeated and refined. In his view, it would be very worthwhile to reexamine the ability of hydrox to scavenge free hydroxyl radicals or alter enzyme functions. "There are several exciting possibilities that should be considered," he wrote in his summary, including new studies to retest whether molecular hydrogen could prevent the formation of cancerous lesions, induced artificially in laboratory mice with certain carcinogenic mixtures that can produce genetic alterations in cells. Fife believes it may be worthwhile to extend this approach and examine whether exposure to hydrogen might scavenge "other free radicals which are found in many industrial environments," perhaps leading the way to minimize the development of occupation-related malignancies. Another area of scientific investigation would be whether hydrogen could play a beneficial role in minimizing the adverse side effects of chemotherapy or radiotherapy. "It should be realized," Fife noted, "that if molecular hydrogen in conjunction with some drug can stop some malignancy, it would add greatly to the armamentarium of the oncologist."

A fascinating aspect of a future hydrogen economy that first attracted scientific attention in the early 1970s is the concept of producing food—single-cell protein—from hydrogen, with the help of certain bacteria and solar or nuclear energy. The idea got its start partly because of NASA's interest in closed-cycle methods of producing sustenance for future astronauts from recycled basic materials during long deep-space missions. Producing some kind of meat substitute refinery-style may be abhorrent to gourmets, advocates of natural food, and a lot of common folk, but it offered the prospect of reducing worries about how to supply an exploding world population with staple primary proteins—a major international policy issue two or three decades ago.

NASA's food-in-space research got underway in the early 1960s, the basic idea being to utilize hydrogen, CO_2, and mineral salts produced by astronauts. Electricity generated by a spacecraft's solar cells would produce hydrogen and oxygen from onboard water supplies in a permanent cycle. Two researchers from the Battelle Memorial Institute, John Foster and John Litchfield, first presented their concept of a continuous-culture machine using hydrogen for protein production aboard a spaceship for deep-space flights in 1964 at a national meeting of the American Institute of Chemical Engineers in Pittsburgh. They said in their summary that the "harvested bacterial substance is high in protein which contains all the essential amino acids." NASA abandoned the project in the 1970s, Litchfield said in a July 1997 telephone interview. Lower animals were not adversely affected, he said, but "primates, humans, definitely were." Studies by nutrition researchers at the University of California at Berkeley showed "pretty clearly" that the material was unsuitable for humans.

As early as 1970, Cesare Marchetti sketched (in a lecture at Cornell University) how hydrogen and certain bacteria can be a link between primary energy and the mechanism of food synthesis:

Chlorophyll is the keystone of the process. Energy from light's photons is accumulated by this phosphor and transferred into ATP, adenosintriphosphate, the universal energy carrier in biological systems. But when an organism oxidizes an energetic substrate, be it sugar or hydrogen, the result is the same. With ATP, the function of chlorophyll is taken up by . . . enzymes. The privileged position of chlorophyll is given by the fact that it is coupled to a primary-primary source of energy, the sun.

Nuclear fission, Marchetti noted, could be substituted for solar energy. (Today, most alternative-energy supporters would stick exclusively with the sun), Marchetti added:

If we find a link between the biosphere and this new source, chlorophyll and agriculture are going to lose their privileged position and the corresponding limitations are likely to fall. Hydrogen can be the link. A certain number of microorganisms are able to use hydrogen oxidation as a source of energy and thrive on a completely inorganic substrate. . . . The energy is used in a quite efficient way to synthesize all sorts of things necessary to build and run the biological machinery-proteins, vitamins, carbohydrates, and so on. The energy conversion efficiency . . . is quite high, 60 to 70 percent in the best cases, 50 percent in an easy routine.

The high multiplication rate of these microorganisms would be a bonus, said Marchetti: "A cattle herd has a weight doubling of two, three years, and a *Hydrogenomonas* 'herd' has a doubling time of two, three hours." Elsewhere, Marchetti noted: "Here then is a prospect of being able to create a food source independent of agriculture (and of oil deposits) linked only to nuclear reactors by a process rather similar to the beer-brewing process." Marchetti made the intellectual leap of relating the energy output of a primary power source—nuclear, solar, whatever—to human food requirements:

. . . a man needs a caloric input of 2500–3000 Kcal per day, corresponding to a rounded mean power of 150 watts. Taking into account all the losses from nuclear energy to hydrogen, and from hydrogen to food synthesized by microorganisms, to have 150 watts 'at the mouth' one should count on roughly 500 watts at the reactor level. This means that a reactor designed to run a power station of current commercial size, say, 1000 Megawatt (electric) . . . and assuming a conversion efficiency in the plant of 40 percent, (this) could be the primary energy source to feed about five million people.

As such a food plant can presumably be packed over a few acres of barren land, one can easily see the consequences of this crude arithmetic.

In Germany, a team headed by Hans-Dieter Schlegel at the Institute for Microbiology of Göttingen University had been investigating the properties of *Hydrogenomonas* bacteria since the 1950s. Schlegel's team also looked into the idea of a closed-loop system for future long-term space missions. But when global food supplies and possible future global food shortages began to loom as international policy issues in the 1960s and the 1970s, Schlegel began to investigate these bacteria more intensively from that point of view. Starting with small culture dishes, the Göttingen team produced protein in "fermenter tanks" up to 200 liters in size. Some of the researchers were concerned mostly with protein production; others were looking at genetic manipulation of the hydrogen-devouring strains in an effort to come up with more efficient mutants. Schlegel explained in a 1971 article:

Basically, molecular hydrogen is not an unusual source of energy for living beings. All aerobic organisms derive the energy necessary for the construction of their cell substance and to maintain their life functions from the reaction between hydrogen and oxygen. Man as well derives his metabolic energy through the slow combustion of hydrogen, in other words, from the so-called Knallgas reaction,[5] although he is not being offered hydrogen in its gaseous state as nourishment, but rather as part of his foodstuffs, weakly bonded to carbon. Metabolic energy is not released through the combustion of the carbon but primarily through the oxidation of the hydrogen contained in foodstuffs. The product of burning hydrogen is water.

Schlegel "harvested" the bacterial cells as a thick suspension. This "soup" was heated for 10 minutes to a temperature of 65°C, activating certain enzymes that broke down the nucleic acids. The "soup" was then run through a separator that strained out the broken-down indigestible nucleic acids, reducing the concentration of these acids from 20–23 percent to 5 percent. The remaining protein-rich cells could be handled quite conventionally; the paste could be dried into a powder. Like milk powder, it could be freeze dried, or it could be pressed into a moist cube like baking yeast. "From the biological point of view," Schlegel said, "we are at the stage that we could do it." But there were risks, such as the less-than-favorable economics and the explosion hazard: "We are working with fermenters with an explosive gas mixture of up to ten liters volume, but I don't dare to go beyond that." Industrial-scale production probably would have meant the use of huge fermenters of up to 400,000 liters, so the volume of the explosive hydrogen-oxygen mixture would be very large.

What about the very idea of eating "synthetic" food, made by bacterial action? In fact, bacteria have been used for centuries to prepare many foodstuffs and medicines, and they are being ingested all the time as cheese or yogurt. Schlegel said that after removal of indigestible or even toxic components—certain storage fats accumulated in the cell and ribonucleic acid[6]—protein quality is high, and the nutritive value is similar to that of the casein found in milk. The protein can be extracted from the bacterial cell mass and processed similarly to the way soybean protein is turned into synthetic meat, sold commercially in health food stores and elsewhere as "textured vegetable protein" (TVP).

In a sense, the idea of producing protein in a laboratory or a factory is an extension of existing industrial practices. Decades ago, British Petroleum set up plants for producing protein from crude oil; other companies too have developed processes using natural gas as a basis for protein. Certain strains of yeast-like microorganisms grow rapidly on an oil base. That discovery, which goes back to the early 1950s, had been exploited industrially to make single-cell protein for animal feed. Similar techniques have been developed to make biodegradable plastics. Techniques for making biologically produced, biodegradable plastics were discovered at the Pasteur Institute in Paris in the 1920s.

In the 1980s, the British chemical industry giant ICI developed a biologically produced biodegradable plastic, trade-named Biopol, that was made essentially by a fermentation process that employed strains of *Hydrogenomenas* (later renamed *Alcaligenes eutrophus*). These bacteria readily transform gaseous mixtures of hydrogen and carbon dioxide and oxygen for use as a slow-release encapsulation material for pharmaceuticals, among other things.

Biopol was never a resounding commercial success. ICI handed it off to a subsidiary, Marlborough Biopolymers; it was then transferred to another subsidiary, Zeneca, which later was sold to Monsanto. For a while, Monsanto supported research at the University of Münster, where a team of scientists headed by Alexander Steinbüchel (a former student of Schlegel's) has been investigating the microbial synthesis and biotechnical production of polyesters since the mid 1980s. One new tack included transferring the process from microbial fermentation to in vitro production in transgenic plants, with the expectation of eventual farm harvesting of these biodegradable plastics. In a 1998 paper,[7] Steinbüchel and Bernd Füchtenbusch wrote that this new production technology will "most probably be feasible in the future," and that versions of these polyesters "might be produced as commodity chemicals for bulk applications, and others as fine chemicals for special applications."

In the end Monsanto didn't share that optimism. It decided to drop the program in late 1998. Michael Berezo, the Monsanto executive who had headed the Biopol program before the shutdown, told me in early 1999 that these materials cost about 10 times as much as comparably performing petroleum based (and not biodegradable) commodity plastics, such as polyesters, some polyolefin, and polystyrenes. Berezo didn't consider biodegradability a "big, positive selling aspect."

A number of scientists and their institutions continue to be intrigued by the prospect of using microbes to produce protein from inorganic materials—especially if the process were to be driven by the sun. The organisms would use hydrogen as an energy source, CO_2 as a carbon source, air as a source of oxygen, and simple minerals from fertilizers to produce a protein-rich, cellular, spongy, whitish material. Paul Weaver, a biochemist and microbiologist working at the National Renewable Energy Laboratory in Golden, Colorado, is investigating this as a sideline to his main research task of making hydrogen fuel in the Department of Energy's Hydrogen Program. Weaver is looking at economical hydrogen production from "any old organic" matter: biomass or coal, for instance, but also municipal solid waste (by far the least expensive raw material) or tree chips. Municipal solid waste is attractive because "you can get a credit" for carting the stuff away and putting it to use again, Weaver says. He figures that hydrogen fuel production could be made cheaper by co-producing a higher-value material, such as biodegradable plastics or high-protein animal feed.

Basically, Weaver gasifies the biomass to make a fuel gas—primarily hydrogen and carbon monoxide. In a second step, the gas would be cleaned up by bacterial action, converting the carbon monoxide component into additional hydrogen in what is known as a water-gas shift reaction. This product gas stream is suitable as a clean hydrogen fuel. The bacteria used for this step are known collectively as photosynthetic bacteria, with individual genus and species names such as *Rhodospirillum rubrum*, *Rhodocyclus gelatinosus*, and *Rhodobacter sulfidophilus*. Although the bacteria are photosynthetic (that is, sustained by sunlight), they perform the shift reaction equally well in either light or darkness for periods of more than a year.

Since the photosynthetic bacteria do the shifting in darkness, the process can run 24 hours a day. On sunny days, the bacteria grow photosynthetically on a portion of the hydrogen and convert it into new cell material: single-cell protein, vitamins, cofactors, essential amino acids, or other nutrients. When nitrogen is kept out of the process, the bacteria cannot make protein, and about 80 percent of the new cell material comes in the form of tiny plastic-containing granules. "When you extract these granules from the cell residue, they are about 98 percent polyester," says Weaver— biodegradable polyester at that. Any of these co-products can be photosynthesized simultaneously with hydrogen, depending on the conditions. There are no waste products, Weaver says. Single-cell protein could be made at about 60 cents per pound.

Bacterial protein is very similar to tofu, Weaver says. "It's also similar to egg white in that protein essentially has no taste. Our taste buds do not respond to protein." Weaver has sampled the stuff: "It mostly dissolves on your tongue. If it were cooked, it probably would be spongy, as are egg white or tofu when denatured." The bacteria are very efficient—essentially 100 percent—in converting the fuel gas components into new cell materials.

Extracting the polyester from bacteria is complicated and a real engineering challenge, Weaver acknowledges. But bacteria cultures grow fast. Weaver says they can double in as little as 90 minutes, but under normal conditions cultures double in mass approximately every 4 hours. With wood chips, Weaver expects, his process would produce biodegradable plastics at $2 per pound—not quite the tenfold reduction that Michael Berezo thought would be needed to be competitive, but a long way toward that goal.

Microbial single-cell protein could have a great impact on world health if generated economically, Weaver believes: "The general goal of industries interested in this product is to produce a high-protein animal feed, which is by far the most expensive component of producing the animal protein that the developed world desires for human consumption." Not only do microbes produce about 65 percent of their cell mass as high-quality protein (Weaver says the amino acid composition is similar to the World Health Organization's "ideal protein" standard); in addition, they contain nearly all essential vitamins and neutriceuticals. Weaver says that German, English, and Japanese companies have been the most interested, and that companies involved in aquaculture (fish farming) are now the most active. These are the areas in which bacteria-driven hydrogen conversion might find a market niche, Weaver believes—not the large-scale production of protein to head off starvation in the developing world, as was envisioned 30 years ago. "Remember," Weaver notes, "the human need for protein is about 20 percent of caloric intake, the rest being carbohydrates and fats. Since single cell proteins (SCPs) are 65 percent protein, supplementation of human food with minor amounts of SCP can improve nutrition significantly. Likewise, the bulk of animal feed is still grasses and grain, but protein needs to be added. The human world is not short of calories. There is a world shortage of protein, however, and more than a billion people show the effects of sub-optimal nutrition, such as in kwashiorkor [a form of malnutrition caused by protein deficiency, especially in young children in the tropics]."

Hydrogen, as an indicator of the presence of water, plays a role in the search for extraterrestrial life. In an October 1992 story about NASA's then newly computerized search for intelligent life in space, *Newsweek* noted that the ongoing search by the planetary researcher Paul Horowitz was focused on hydrogen's frequency of 1420 megahertz. "This simplest and most abundant atom in the universe," *Newsweek* explained, "vibrates at a frequency of 1,420,405,751 cycles a second, a frequency that Horowitz says would make sense as a meeting place in the vast radio spectrum."[8]

NASA stopped funding the radio search as an economy measure in the early 1990s, but Horowitz and his team are continuing to scan the heavens with private funding. At the time of the appearance of the aforementioned *Newsweek* article (1992), the Megachannel Extra Terrestrial Assay (META), largely paid for by the filmmaker Steven Spielberg, was scanning a huge number of frequencies while sticking fairly close to that of hydrogen. After searching 60 trillion channels over 5 years, META found 37 "candidate events," as one web site put it with sober restraint. Unfortunately, not one of them was detected again.

In 1995, META became BETA (Billion-channel ExtraTerrestrial Assay), a faster, more discriminating, automatic sweep of a billion channels at a time, covering the full 1400–1700-megahertz band of "water hole" frequencies. Every candidate channel is checked twice—first in the east, then in the west. If a channel looks promising, the antenna is programmed to break off its survey and lock onto the candidate signal, sweeping on and off the exact wavelength to capture other nearby frequencies.

And when in early 2001 one clicked on the highlighted question "Have we found aliens?" at Harvard's BETA web site, an ominously all-black screen popped up, with tiny letters squeezed into the upper left corner: "No, not yet."

Project Phoenix, at the SETI Institute in Mountain View, California, is a similar effort. It looks at a wider spectrum of signals, between 1000 and 3000 megahertz, in very narrow (1 hertz) channel segments, or about 2 billion channels for each target star. Project Phoenix doesn't cover the whole sky; it listens to space around "nearby" (no more than 200 light years away) sun-like stars considered most likely to have evolved long-lived planets that may be capable of supporting life. By mid 1999, Project Phoenix had examined about 400 of its 1000 target stars. "No clearly extraterrestrial transmissions" have been found, according to the Phoenix web site.

11

Safety: The Hindenburg Syndrome, or "Don't Paint Your Dirigible with Rocket Fuel"

On March 27, 1977, two 747 jumbo jets crashed into each other on a foggy runway on Tenerife in the Canary Islands when a KLM plane struck a Pan American plane just as it was lifting off. Of the 644 passengers aboard the two planes, 583 perished in the worst disaster in the history of civil aviation.

G. Daniel Brewer, the former manager of Lockheed's hydrogen program and probably the US aircraft industry's most outspoken advocate of liquid hydrogen as a fuel, maintains that the carnage could have been much reduced if the two planes had been fueled with liquid hydrogen instead of conventional jet fuel. After analyzing information that became available shortly after the disaster, Brewer concluded that many of the deaths were probably due to the kerosene-fueled fire that raged for about 10½ hours. "If both aircraft had been fueled with liquid hydrogen, there is a reasonable possibility that many lives could have been saved," Brewer told an audience of experts 6 weeks after the tragedy at a hydrogen symposium at the European Community's Joint Research Center in Ispra, Italy. Some 25 or 30 persons probably would have been killed outright by the direct impact, Brewer reasoned. Fire presumably would have broken out, but because of hydrogen's different burn characteristics many passengers could have been rescued. Brewer's reasoning went as follows:

• The fuel-fed portion of the fire would have lasted only a minute or two, because hydrogen is so volatile and because it is highly unlikely that both LH$_2$ tanks in both aircraft would have been ruptured.

• The fire would have been confined to a relatively small area. The LH$_2$ would have vaporized and dispersed before it could have spread widely.[1] Also, a much smaller portion of the total fuel carried by both planes would presumably have been spilled.

- Radiation, or radiated heat from the fire, would have been significantly less, so that only those persons and that part of the structure directly above or next to the flames would have been burned.

- The absence of smoke from the burning hydrogen fuel might have saved some lives. However, smoke from other incendiary material in the plane probably would have negated this advantage for hydrogen.

Whatever the merits of Brewer's ideas, one cannot help but compare the Tenerife disaster, which faded fairly quickly from the public's consciousness, with the *Hindenburg* dirigible accident of 1937, in which 36 people died—35 passengers and crew members and one member of the ground crew. The disaster received saturation coverage in the news media, and many advocates of hydrogen as a fuel fear that it made "hydrogen" a negatively charged word in the popular consciousness.

The *Hindenburg* did not explode, as is often believed; it burned. Furthermore, its sister ship, the *Graf Zeppelin*, made regular scheduled transatlantic crossings from 1928 until its retirement in 1937 without a mishap. And before the disaster, the *Hindenburg* had successfully completed ten round trips between the United States and Europe.

Although sabotage was initially suspected as the cause of the conflagration, many have believed for a long time it was more likely that electrostatic charges present in the atmosphere after more than an hour of thunderstorms and rain ignited the hydrogen that was being vented as the crew was trying to tie the big dirigible to its mooring tower.

It should also be remembered that the *Hindenburg* and its sister dirigibles were designed to be inflated with non-flammable helium. Hydrogen had to be substituted when the United States, the sole source, refused to sell helium to the Germans in the aftermath of World War I.

Almost exactly 60 years after the disaster, the widely respected NASA veteran Addison Bain uncovered impressive evidence that persuasively exonerated hydrogen as the primary cause of the disaster and that showed that static electricity and the presence of highly inflammable materials in the airship's skin were to blame. Bain, the former manager of the hydrogen program at the Kennedy Space Center, presented the findings of his decade-long investigation in the spring of 1997 at the annual meeting of the National Hydrogen Association. His research had included exhaustive analyses of surviving fragments of the *Hindenburg*'s cotton-based covering performed in NASA's Materials Science Laboratories. His conclusions, first published in *Air & Space Magazine*, were reported in a large article in the science section of the *New York Times*. They generated a cover story in *Popular Science* and several TV documentaries. In his NHA keynote address, Bain said he had discovered that the use of cotton fabric and a doping process that had involved aluminized cellulose acetate butyrate and iron oxide had combined to produce a material likely to burst into flames with a minimum of incendiary incentive. "The Space Shuttle's solid rocket boosters use powdered aluminum as the fuel and iron oxide as a burning-rate catalyst," Bain noted drily.

Static electricity in the air (witnesses reported seeing blue discharges on top and in the back of the ship, near the point where the flames first occurred, moments before the fire) most likely provided the spark that set off the conflagration, Bain reasoned. "Atmospheric and airship conditions at Lakehurst were conducive to formation of a significant electrostatic activity on the airship," he noted. "Hydrogen naturally contributed to the fire," said Bain. "It is a fuel, and fuels must be flammable. But the airship envelope was sufficiently combustible so that it could have burned even if the dirigible had been filled with an inert gas like helium. It was really a cellulose fire, flavored with some hydrogen." The clincher of his presentation was his final slide, which showed a US Navy dirigible on fire in July 1956 at a Naval Air Station in Georgia, in the rain, with firefighting equipment nearby. To the casual observer it looked very much like the *Hindenburg*, with the envelope burning rapidly. But this airship had been buoyed by inert, non-flammable helium—not hydrogen. "The moral of the story," Bain concluded, is "don't paint your airship with rocket fuel."

Safety was on the international hydrogen community's agenda from the start, and in the 1990s hydrogen safety research was underway in laboratories in the United States, in Europe, in Japan, and probably elsewhere.

In 1994, some of Dan Brewer's ideas about the behavior of liquid hydrogen were confirmed by an experiment in which copious amounts of LH_2 were intentionally spilled and ignited in a row of abandoned military barracks near Berlin that had once housed Soviet troops. Scientists from Germany's BAM (Bundesanstalt für Materialforschung und -prüfung—that is, Federal Institute for Materials Research and Testing), from the German branch of the Battelle Memorial Institute, and from the Jülich Research Center, intentionally spilled some 650 liters of LH_2 and about 100 kilograms of liquid petroleum gas (LPG) (a mixture of liquid propane and butane). Each fuel formed big puddles on the ground, on a metal sheet, and on a water surface in a lane between two barracks, according to a paper presented at 1994 International Cryogenic Engineering Conference, in Genoa. The authors said that the series of tests, conducted over 4 days and employing some twenty sensors for the hydrogen test alone and five video cameras, represented the first attempt to find out what would happen if such spills were to occur in a residential environment. Previous spill experiments had yielded important data about vaporization and cloud formation, the authors said, but most of them had been conducted "in an open area under conditions that are not all typical for a real accident."

What happened? Not much. Liquid hydrogen didn't exhibit any striking differences from liquid natural gas (LNG), LPG, and other conventional fuels. "What we have demonstrated is that even in the case of an accident, LH_2 is not more dangerous than these other fuels," the researchers reported. In fact, it may be even safer in some respects because it evaporates quickly instead of spreading on the ground. A BAM press release said it was virtually certain that "liquid hydrogen is not any more dangerous in such an accident scenario than the well-known propane." According to the release, "the hydrogen cloud that had been created rose very quickly and dispersed, while propane initially collected on the ground but was subsequently dispersed by air turbulence."

As early as 1956, Arthur D. Little and Lockheed ran tests to get an idea of what hazards existed in handling the large quantities of hydrogen that were expected to be used in the US space program, and of what procedures would be required. The Arthur D. Little researchers spilled and ignited as much as 5000 gallons of LH_2 in an open space, but there were no explosions.

In a 1967 paper, A. A. Du Pont, a researcher for the Garrett Corporation of Los Angeles, reported that the "use of hydrogen in normal aircraft operation can be as safe or safer than the use of fuel," although "questions arise as to the relative hazard of hydrogen fuel in case of a serious accident where the airplane is partially destroyed and the fuel is spilled." Du Pont said it is "very difficult" to obtain an explosive air-hydrogen mixture even in a confined space—a claim that others disputed. He also asserted that a hydrogen fire is "very much better from a safety standpoint." "A spill of cryogenic hydrogen," he continued, "boils furiously on contact with the relatively warm ground, so the area of spilled fuel is therefore confined whereas the spill of hydrocarbon fuel spreads along the ground and covers a considerable area." In an accident in which fuel is spilled and ignited, a kerosene fire covers a much larger area than a hydrogen blaze, Du Pont said. Overall, he concluded, "it appears, after objectively evaluating the properties of the two fuels and the probable size of the fuel loads involved, that a passenger's chances of walking away from a crash of a hydrogen-fueled airplane are as good or better than they would be with a kerosene-fueled airplane."

Du Pont may have been a bit overoptimistic. A 1976 Stanford Research Institute study pointed out that comparing hydrogen against conventional fuels in regard to safety yields "no clear cut answers." Rather, the study maintained, "since the physical and chemical properties of hydrogen are quite different from commonly encountered substances, the hazards associated with hydrogen appear to be far more influenced by circumstances than for other fuels . . . blanket statements about the safety of hydrogen relative to other substances are generally misleading." The SRI study suggested that "public perceptions of the safety of hydrogen may be one of the major obstacles to a transition to a hydrogen economy."

In the mid 1970s, researchers at Wright-Patterson Air Force Base fired armor-piercing incendiary and fragment simulator bullets into styrofoam-lined aluminum containers, some filled with LH_2 and others with petroleum-based JP-4 jet fuel. They also simulated lightning strikes into the containers. These experiments, they reported, showed LH_2 to be "more forgiving" than kerosene. As described by Jack Lippert at the 1976 Miami Beach hydrogen conference, the results indicated that the incendiary weapons ignited but did not detonate the LH_2. The hydrogen fire was "less severe" and "expired more quickly" than a comparable JP-4 fire, even though the total heat content of the hydrogen sample was twice that of the JP-4 sample.

Blasting away at the sample containers with the nonexplosive fragmentation bullets demonstrated that the LH_2 experienced little "hydraulic ram effect" (an internal pressure wave building up in the liquid that could rip open the container or tank and cause damage to surrounding sheet metal). The nonexplosive bullets also did not ignite the LH_2, which simply poured out through the entry and exit holes.

When a similar styrofoam-covered container filled with JP-4 was shot at, the bullet caused overpressure that forced the kerosene through the lid and ripped a fairly large exit hole. Lippert attributed the difference in behavior to the fact that liquid hydrogen is only about one-tenth as dense as the JP-4. A bullet creates much more impact and shock wave in a high-density fluid than in a low-density one.

To simulate lightning strikes, the researchers used a 6-million-volt generator to shoot big arcs into the liquid-hydrogen containers. Again, there was ignition but no explosion. "The data indicate," Lippert concluded, "that the hazards associated with LH_2 utilization in combat aircraft may be less severe than those with JP-4, and "therefore it is recommended that LH_2 should not be disregarded as future alternate fuel for military as well as commercial aircraft."

Some hydrogen supporters feel that, in view of the complexity of rocket technology, NASA's experience may not be very relevant to the everyday safety problems likely to be posed by a future civilian hydrogen economy. Still, it is noteworthy that NASA has used and handled stupendous amounts of LH_2, most of it hauled by barge and tanker-trailer over hundreds of miles to Cape Canaveral and other sites.

In 1974, a paper reviewing 96 hydrogen incidents and accidents was presented by NASA researcher Paul Ordin at the Ninth Intersociety Energy Conversion Engineering Conference in San Francisco. Ordin reported that NASA's tanker trailers had hauled more than 16 million gallons of LH_2 for the Apollo-Saturn program alone. He described 17 mishaps involving tanker trailers carrying between 3000 and 16,000 gallons. Twelve of these had occurred during offloading; five involving highway accidents. Taken together, the results seemed to indicate that LH_2 was fairly forgiving in potential high-risk situations.

In three of the five highway accidents, the tractor was heavily damaged but the trailer remained intact; in the other two, the truck was not damaged. In the one case in which LH_2 spilled onto the ground, there was no ignition. Any of these accidents might have caused a spectacular blaze had the load been gasoline or kerosene. In one case, a tractor-trailer that had entered a curve too fast rolled 40 feet down an embankment. The tractor rolled over once, the trailer one and a half times. The tractor "was a complete loss," but the trailer was little damaged, with the shell in satisfactory shape and venting normally. In another mishap, a hydrogen carrier jackknifed on a wet road in a strong wind after braking, swung around 180°, and slid onto the shoulder. The trailer unhitched, went into a ditch, and slid about 80 feet on its side. The safety disks ruptured as designed, and about an hour later the LH_2 had safely vented into the air. In a third case, a tractor-trailer returning from fuel delivery but still containing gaseous hydrogen and traveling at about 50–55 miles per hour, was hit in rainy weather by a truck that spun in front of it. The tractor was demolished, but the trailer did not suffer any leaks, and the vacuum was not broken.

Hydrogen was released in 80 of the 96 incidents; fire broke out in 61. Though this may sound like a high percentage, the researchers found it more significant that in 25 cases there was no fire.

Although NASA's experience "offers encouragement about the safe use of liquid hydrogen" and should serve as a point of departure for developing future safety procedures, the SRI study concluded that NASA's record was "inadequate to serve as a basis for decision-making" in regard to a more general hydrogen economy. The NASA personnel who handled hydrogen were specially chosen and trained, the use of hydrogen in the space effort was restricted to "rigorously controlled" outdoor environments to limit explosion hazards, "extreme concern about the reliability of systems" was dominant in the entire space program, and NASA's hydrogen safety procedures involved "expensive, often unique" hardware.

Another departure point for assessing hydrogen's safety was provided by the experience of the world's first long-distance pipeline for transporting hydrogen to various chemical plants, operated by Chemische Werke Huels (CWH) in Germany. This 130-mile network is still in operation. In considering its safety record, it is important to remember that the pipes are very small in diameter, nowhere near the diameter expected to be needed for future energy pipelines, and that they operate at pressures that are very low in comparison to the thousands of pounds per square inch demanded for efficient energy transport. Also, in some instances the hydrogen gas being piped is not very pure (about 95 percent or even less); this bears on leakage, embrittlement, and flammability, since lower purity reduces the hazards somewhat.

Christian Isting, a Huels executive, said in a paper delivered at a 1974 European energy symposium that "compressed hydrogen frequently ignites on expansion." The reasons for this are not clearly understood yet, but they probably have to do with electrostatic charges caused by dust particles in the air. Pressurized hydrogen would almost always ignite if vented through a seldom-used line, Isting reported, but ignition could be generally avoided if initially only small amounts were sent through the venting line, and amounts and flow speed were increased gradually. Large amounts can be vented to the atmosphere without ignition except during thunderstorms. "During the many years of operation of the integrated pipeline network of CWH," Isting concluded, "explosions have not occurred. Either the hydrogen will ignite immediately on escape or, if no ignition takes place, an explosive mixture cannot form near the ground because of its low density. Explosions or even detonations in the pipeline need not be feared as explosive gas-air mixtures cannot build up because of the pressure prevailing in the line."

To understand the safety implications of hydrogen more fully, it is useful to look at hydrogen's physical properties, as explained by the 1976 SRI study:

1. Liquid hydrogen is a very cold material, –423°F (–252.8°C). Contact with human body tissue can result in severe burns, destroying tissue almost like the burn from a flame.
2. As a gas, hydrogen diffuses very quickly into other areas, such as air. It has a very low density, meaning it rises very quickly through the air.
3. Mixed with air, hydrogen burns over a much wider range than methane or gasoline, for instance. A mixture of as low as 4 percent of hydrogen in air and as high as 74 percent will burn. The corresponding ranges for methane are 5.3 to 15 percent, and for gasoline it is a very narrow range from 1.5 to 7.6 percent. Jet fuel flammability range is even narrower, from 0.8 to 5.6 percent.
4. When confined in a completely enclosed space—a room, or a tank—hydrogen can be detonated, i.e., exploded over a wide range of concentrations, ranging from 18 to 59 percent (by volume) in air. Methane explodes only in concentrations ranging from 6.3 to 14 percent, and gasoline and jet fuel detonate in a range from 1.1 to 3.3 percent.
5. It takes very little energy to ignite a hydrogen flame, about 20 micro-joules (MJ). Methane requires about 12 times as much energy, about 290 MJ, to set off

a burn, and gasoline needs about 240 MJ.[2] However, out in the open without confinement, it is almost impossible to bring hydrogen to an explosion with a spark heat or flame. Hydrogen-air mixtures can be detonated only with a suitable initiator, such as a heavy blasting cap.

6. The hydrogen flame is almost invisible in daylight (unless some special colorant is added), and it travels much faster than a methane flame. A hydrogen flame shoots upward at a rate of 2.75 meters (9 feet) per second, while methane and gasoline burn much more slowly at 0.37 meters (1.2 feet) per second.

7. Unlike kerosene or gasoline flames, hydrogen flames radiate very little energy, meaning heat is not felt at a distance.[3]

Clarifying, the study added the following:

While the invisibility of the flame makes the fire difficult to locate and fight, firefighters can get very close to the flame (assuming they know where it is) without injury. Since it is also difficult to feel warmth from the flame, a person can easily move right into the flame and be burned; but surrounding objects do not heat up and ignite unless touched by the flame directly. In normal fires, one of the major causes of fire spreading and injury is the large amount of energy radiated by oxidizing carbon atoms. Thus, the lack of flame luminosity in hydrogen fires can be both a help and a hindrance.

The energy radiated from fires above pools of liquid hydrogen and liquid methane of the same area and volume would be nearly identical because the lower emissivity of hydrogen is offset by the more rapid rate of combustion. Of course, this also means that the available fuel is more quickly exhausted in the hydrogen fire. However, owing to its rapid rate of evaporation, pools of LH_2 are not likely to form except after a very large spill. A liquid-hydrogen fueled airplane that crashed on takeoff would probably result in such a spill.

Hydrogen's wide range of flammability and low ignition energy is offset by its great tendency to disperse from the scene of a leak or a spill because of its buoyancy and high rate of diffusion. . . . When gasoline is spilled, vapors heavier than air spread in a wide layer near the ground. This greatly increases the hazardous area. Thus, the tendency of hydrogen to disperse much more rapidly than other fuels, even gaseous methane, is a large point in its favor.

Since the early investigations, a great deal more has become known about hydrogen, and detailed research into safety fundamentals is continuing. At the 1996 World Hydrogen Energy Conference in Stuttgart, for example, more than 30 of the 375 papers and posters dealt with materials and safety. Papers covered topics as esoteric as "Fracture mechanical behaviour of the steel 15 MNNI 6 3 in high-pressure hydrogen gas with admixtures of oxygen," "Turbulent combustion and detonation processes in hydrogen-air mixtures," and "Hydrogen-influenced mechanical behaviour of austenitic steels and nickel base alloys." But there were also reports on subjects that seemed closer to consumer reality, such as "Safety device for hydrogen appliances," "Safety considerations in the design of hydrogen-powered vehicles," and "Experimental examinations on the worst-case behaviour of LH$_2$/LNG tanks for passenger cars."

In 1993, a more general safety study commissioned by Germany's parliament, the Bundestag, concluded that the risks posed by the wide use of hydrogen are likely to be relatively low and could be managed fairly easily. The study, undertaken by the Bundestag's Büro für Technikfolgen-Abschätzung (Office for the Assessment of Technology Consequences), complemented the work of an earlier national inquiry commission, the so-called Enquete Kommission, which had grappled with a host of energy and environmental issues, including prospects and consequences of development strategies leading to a solar hydrogen economy. "The technical risks in all components of a hydrogen energy system, from production to utilization, are, in principle, regarded as controllable," said the five-page executive summary. "The main risks appear to be in the middle of the hydrogen chain concept, both during storage and during distribution of hydrogen in both gaseous and liquid form." The study listed hydrogen's advantages in various accident situations, including release, vaporization, cloud formation, fire, and lower thermal emissivity. It listed as a disadvantage peak pressures during accidental combustion of hydrogen-air mixtures, especially in closed or partially closed rooms. (In view of unresolved questions about materials and embrittlement, the risk involved in shipping gaseous hydrogen in high-pressure pipelines could not be assessed in any meaningful way.) The study saw the risk associated with the handling of hydrogen by untrained individuals (i.e., ordinary consumers) as requiring technical safety measures, such as "automated event sequences" and self-adjusting gas sensors in closed rooms linked to ventilator systems.

One of the best-known American hydrogen safety researchers is Michael Swain, an associate professor of mechanical engineering at the University of Miami at Coral Gables. For many years, Swain (whose involvement with hydrogen goes back to the early 1970s, when he and a colleague converted a Toyota station wagon to hydrogen) has been looking systematically at safety issues. Summarizing past work at the 1998 DoE Hydrogen Program review in Alexandria, Virginia, Swain said that he and his colleagues had investigated hydrogen safety issues associated with automobiles, buildings, pipes, residential fuel lines and pipelines, hydrogen facilities of all types, ventilation systems, accidental combustion during hydrogen production, flame arrestors, and hydrogen permeability.

At the US Department of Energy's 1996 Hydrogen Program review meeting in Miami, Swain reported that the energy leakage rate of hydrogen was less than that of natural gas or liquid petroleum gas in similar-sized leaks with the same line pressure, even if the line pressure was increased to produce the same energy flow rate as for natural gas or LPG. Swain compared the behavior and the potential danger of clouds of hydrogen with those of LPG via computer model analysis as well as experiments. He found that in most cases LPG clouds were more dangerous than hydrogen clouds. This, Swain said, was due to different gas densities: the movement of a gas cloud in the air was dominated by the cloud's density, diffusion playing a secondary role. In examining various scenarios that could lead to accidents, Swain found that LPG, with its higher density, usually created more danger than hydrogen.

In yet another analysis, Swain found plastic pipes to be about twice as permeable to hydrogen as to natural gas. Even so, the risks were "still negligibly small from a safety standpoint." Swain also found that "lean hydrogen flames are not visible in the presence of other light sources, but rich hydrogen flames are quite visible," and that "the addition of small quantities of gaseous hydrocarbons increases the flame visibility of hydrogen substantially."

At the DoE's 1995 Hydrogen Program review meeting in Coral Gables, Swain had presented a safety analysis of what might happen when high-pressure hydrogen-storage containers of a type that might be used in a future medium-size hydrogen-powered car were to be punctured in an accident. He had found that both hydrogen leaks and natural gas leaks produce cones of gas mixtures that can be ignited. Lighting the hydrogen-air cone produces more overpressure than lighting the corresponding natural gas–air cone, but the overpressure is still quite low, Swain said—about the same as "overpressures (generated) by inflating an 11-inch child's balloon until it burst." Venting low-density gaseous fuel upward reduces the likelihood of contact with ignition sources during an accident, and venting also reduces the likelihood that gases will be trapped inside the vehicle, he added. Storage of compressed gases in many small containers rather than a few large ones can reduce the amount of gas released in an accident.

Since hydrogen doesn't smell and can't be seen, detecting its buildup inside enclosed spaces like a car's passenger compartment is an important safety issue. At that DoE's 1995 Hydrogen Program review meeting, David Benson and C. Edwin Tracy of the National Renewable Energy Laboratory reported on the development of a fiber-optic detector for detecting hydrogen leaks that may be cheaper to produce than conventional devices (which usually need electrical wiring for their operation and which may be too expensive for mass use in cost-sensitive consumer goods such as automobiles).

The Benson-Tracy detector essentially consists of a reversible thin-film chemochromatic sensor attached to the end of a polymer optical fiber. Exposing that film (a layer of transparent amorphous tungsten oxide covered by a very thin reflective palladium layer) to hydrogen causes it to darken, and the difference can be measured by a reflected light detector. This approach had been demonstrated by Japanese scientists in the early 1980s, Benson said. The project was a spinoff of 10 years of investigations of using tungsten oxide films as electro-optically active electrochromic window coatings. As described by Benson, the main challenge (other than lowering costs and achieving rapid response and sufficient selectivity) was to avoid false readings from whiffs of the exhaust of passing cars.

In the late 1980s, the international hydrogen community began to recognize that hydrogen safety will have to be dealt with at the international level if hydrogen is ever going to become a global energy carrier. In addition, it had become clear to many that hydrogen safety is qualitatively different from the safety of other fuels, and that international standards would have to be established. This led, in early 1990, to the establishment of a new "Hydrogen Energy" technical committee as part of the International Organization for Standardization, the Geneva-based international body that administers all types of technical standards worldwide. The new group, known as Technical Committee 197, was organized largely through the efforts of a Swiss hydrogen supporter, Gustav Grob, who served as the first chairman.

Its stated purpose was "standardization in the field of systems and devices for the production, storage, transport, measurement and use of hydrogen for energy utilization." At the first session, held in Zurich, the organizing committee called for the establishment of several subcommittees, such as Storage, Transport, Production, and Safety and Safety Devices.

Switzerland relinquished the TC 197 secretariat in 1993, and Canada took over. Since then, the committee and its various working groups and subcommittees has met several times, usually in conjunction with one of the biannual World Hydrogen Energy Conferences. By 1995, the group counted 13 countries as permanent and 11 more as observer members. But at the group's June 1995 meeting in Quebec City it became evident that TC 197 was still a work in progress, with no real power to enforce hydrogen standards internationally. "Hydrogen is the same element in all countries, and safety requirements should be the same everywhere," a report of the meeting said. "The hope is that by working together member countries can optimize the materials of construction and the methods of manufacturing, testing, and operation of hydrogen systems. . . . ISO/TC 197 strives to be recognized as the preeminent entity responsible for developing international standards, codes, and guidelines in the field of hydrogen energy technologies." Three working groups were developing specifications for liquid-hydrogen fueling systems, for liquid-hydrogen tanks, and for hydrogen as fuel. The development of standards for hydrogen as an aircraft fuel was added as a new task.

In the United States, hydrogen as a fuel is confronted by a plethora of safety rules and regulations, almost none of them directly relevant to energy use. That this gap will have to be addressed somehow was a major theme of a January 1995 safety workshop organized by the National Hydrogen Association. "There has been a big void in that area," said one of the organizers afterward. Another participant, William Summers of the Westinghouse Savannah River Company,[4] said that safety might be the major issue in regard to the expanded use of hydrogen, and that it was essential to make all stakeholders—the general public, officials, regulators, equipment manufacturers, the insurance industry, investors—comfortable with hydrogen as a fuel. Citing the results of a quick, small, informal survey, Summers found that public knowledge of hydrogen was pretty abysmal. Almost none of the 22 individuals queried knew that hydrogen could be used as a fuel, and all of them said they didn't see any advantages. All felt the stuff was dangerous, and most of them believed exhaust from a hydrogen engine would be toxic.

There is no technical connection between the hydrogen bomb and hydrogen fuel. Hydrogen fuel represents chemical energy and combustion/burning processes; the hydrogen bomb works at the atomic level via principles of nuclear physics. But, as Dan Brewer noted in his Ispra paper, the linkage provided by the word "is apparently enough to stir the imagination of the public and excite fear and suspicion of the fuel."

12

The Next 100 Years

Icy winds whipped snow and sleet through the streets of Reykjavik on February 17, 1999, whistling past the entrance of the luxurious Grand Hotel. Inside, in a conference room, some four dozen people—international business executives, energy technologists, environmentalists, Icelandic parliamentarians and politicians, representatives of the local media—had assembled for a press conference at which representatives of the Icelandic government, of a newly formed Icelandic business consortium, and of three multinational corporations—DaimlerChrysler, the Royal Dutch/Shell Group, and Norsk Hydro—were to announce a million-dollar joint venture to investigate hydrogen's potential as a clean substitute for fossil fuels in Iceland.

The initial six Icelandic participants hold the controlling 51.01 percent of the joint venture. They are the country's fertilizer plant (which for decades has been making hydrogen for ammonia production via electrolysis), the University of Iceland, the Reykjanes Geothermal Power Plant, the Icelandic Research Institute, the New Business Venture Fund, and the Reykjavik Municipal Power Company.[1] The partners were telling the world they were banding together to set up shop for the planet's first laboratory for a future Hydrogen Economy.

Hailing the project, Iceland's minister for environmental affairs, Gudmundur Bjarnason, urged the partners to "cooperate in a fruitful way." "Perhaps their work will even mark our everyday life in time to come," he said. The government, he declared, supported the project "in order to diversify the economy and lay the foundation for higher living standards in the future." He added: "The government of Iceland welcomes the establishment of this company and considers that the choice of location for this project is an acknowledgement of Iceland's distinctive status and long-term potential."

The Reykjavik announcement was a milestone in a process that had begun years earlier. The idea of trying out a prototype hydrogen economy on a large scale on an island such as Iceland had been around for decades. (In the 1970s and the 1980s, Hawaii had been regarded as a good candidate.) A paper presented at the 1992 World Hydrogen Energy Conference by B. Arnason, T. Sigfusson, and V. Jonsson had laid out the specifics for the undertaking: Iceland's economically exploitable hydropower potential is about 30 terawatt-hours per year. (Its geothermal potential is much larger, about 200 TWh per year.) The large fishing fleet, which consumes about 230,000 tons of oil per year (about 12 percent of total consumption, and almost a third of all fossil-fuel imports) could serve as a demonstration project, the oil being replaced by about 70,000 tons of hydrogen per year. And hydrogen energy could be a viable future export to Europe.

Things began to move in earnest in May 1998 with a low-key, largely unreported exploratory visit by a team from Daimler-Benz and Ballard Power Systems. The immediate result was a memorandum of understanding in which the parties agreed to cooperate in devising a road map to turn Iceland (with a population of about 265,000) into a laboratory for hydrogen energy technology and, within 15 or 20 years, into a hydrogen economy. If things work out, eventually all of Iceland's cars and buses will run on either hydrogen or methanol, and the fishing fleet will be completely converted to fuel cell power, fueled by hydrogen derived from hydropower and geothermal energy.

Since the very beginning of the interest in hydrogen as energy, a vexing question had been how to actually get a hydrogen economy going. Endless debates characterized the international hydrogen scene: What should come first, economical hydrogen production or hydrogen-burning machinery? In the 1970s, the heyday of early hydrogen enthusiasm, one assumption was that hydrogen would somehow blend painlessly into the energy market. It was believed, for instance, that hydrogen could be mixed in ever-larger proportions into pipeline gas, eventually replacing natural gas altogether.

However, it soon became clear that, even if problems like pipeline embrittlement or availability of high-pressure compressors were solved, simple things such as metering the gas flow and energy content would present huge difficulties. (At different hydrogen concentrations, meters would give readings that would not reflect the true energy content of the various mixture ratios.) Similarly, early hydrogen enthusiasts believed that hydrogen produced from water would find a ready market in ammonia production, for instance, as an entering wedge via the chemical industry. That didn't happen, and in the 1990s hydrogen produced via steam reforming from natural gas was still cheaper than hydrogen produced via water electrolysis.

One view that developed in the 1970s among some hydrogen theoreticians was that the entire pattern of energy use, including the adoption of hydrogen, would evolve in three long stages. According to this scenario, in the first stage, coal in all its forms—mineral, liquid, gas—would gradually replace oil in 20 years or so, while light-water nuclear reactors would come into wider use and first trials of breeder and high-temperature reactors would be conducted. Work on hydrogen would continue, but it would not make any significant inroads. In the second stage, a 50-year period, a hydrogen economy "in the widest sense" would be phased in. Oil and natural gas would be largely eliminated as primary sources, replaced by solar and nuclear energy (high-temperature and breeder reactors). For cars, methanol rather than hydrogen would most likely be the best-suited fuel. Beginning some time around 2040, the third stage—the golden age of environmentally clean, planet-wide energy systems—would begin. Ocean mining would flourish, primary energy would come from solar sources of all types, and perhaps clean nuclear fusion power would come into the picture. Energy might be generated at a handful of very large sites far removed from population centers, the generated energy being transported to population centers either as electricity (through long-distance, low-resistance cryogenic cables), as hydrogen gas (in pipelines), or as cryogenic liquid (in huge tankers).[2]

So far, however, oil is still very much around, and the oil industry seems bigger than ever (although it is beginning to fray at the edges, as evidenced by Shell's decision to set up a hydrogen division). Coal is still the world's biggest energy source, although obviously not in the all-important transportation sector. Nuclear fission, while important in some countries such as France, is on the decline, and breeders and high-temperature reactors haven't gotten off the ground.

In the late 1990s, hydrogen was making glacially slow progress; widespread hydrogen use was still nowhere in sight. This was to be expected, at least according to the energy theorist and systems analyst Cesare Marchetti. The notion that phasing in hydrogen as a major energy carrier is an inherently long-term (50–100 years) proposition is grounded in Marchetti's 1970s investigations of the rate at which new energy systems—indeed, new major technologies of any type—replace older ones.

Marchetti (still one of the most provocative long-range thinkers in the field) started from the "somehow iconoclastic hypotheses that the different primary energy sources are commodities competing for a market, like different brands of soap, or different processes to make steel, and that the rules of the game may, after all be the same," as he put it in a November 1974 lecture in Moscow. He drew largely on the work of two American scientists, J. C. Fisher and R. H. Fry, who analyzed substitution rates of new technologies and who attempted to forecast the rate of diffusion of new technologies for General Electric.

Using certain basic data and formulas, Marchetti found that he could predict market shares decades in advance. Marchetti produced some amazing graphs that showed the market-penetration curves for various pairs of competing products or technological processes over periods ranging from 60 to 80 years—for example, the replacement of the open-hearth method of making steel by the Bessemer method, the substitution of sulfate turpentine for natural turpentine, and the replacement of water-based paint by oil-based paint. The astounding aspect of these graphs was that the equations describing market penetration worked well as a forecasting tool spanning decades. The curves derived from these formulas matched the historical data with "extraordinary precision," according to Marchetti.[3] The implication was that if market penetration proceeded according to some mysterious underlying law—some force uninfluenced by legislation, short-term economics, or price—it should be possible to predict with reasonable certainty to what extent a fundamentally new product, technology, or energy system would replace its precursors.

Marchetti decided to test the theory by trying to predict oil's percentage of the US energy market, using data from four decades earlier. As he told his Moscow audience, the results were astonishing:

I took the data for the US from 1930 to 1935 and tried to forecast oil coverage of the US market up to 1970. The predicted values, even for the saturation period, fit the statistical data better than one percent which, after all, is the minimum error that can be expected from this kind of statistics. This means that the contribution of oil to the US energy budget, e.g. in 1965, was completely predetermined 30 years before, with the only assumption that a new primary source of energy, e.g., nuclear, was not going to play a major role in the meantime. As the history of substitutions shows, however, the time a new source takes to make some inroads in the market is very long indeed—about a hundred years to become dominant, starting from scratch.

In 1976, at a microbial energy conversion seminar in Göttingen, Marchetti observed that the "takeover time" (the time it takes a new technology to go from 1 percent of the market to 99 percent) was about 50 years for the United States—an observation that presumably would be valid for hydrogen. "The extreme stability of the functions over very long periods of time, including wars, depressions, economic miracles, and a perceived acceleration of knowledge and 'progress'" do not detract from the validity of this phenomenon, Marchetti said. "My feeling is that all this is linked to learning processes at the societal and individual level which evolve very slowly and perhaps were the same 1000 years ago." "The takeover times," Marchetti observed, "are on the order of a century, so there is not much purpose in darting ahead. . . . Meaningful influences on the system can be obtained only through really long-term thinking and planning. . . . In fact, the spreading of a new technology always follows the rule of penetrating first small favorable eco-niches, acquiring force and momentum for the next step." He illustrated that thesis with a thumbnail description of how crude oil wormed its way into the world's energy capillaries: "The oil industry was not born as a competitor to coal, too big an industry and too well entrenched to fight against, but as a welcome substitute for dwindling resources of whale oil for illumination. Being small and aggressive, through the development of a superb chemistry, it was able to conquer the new, very specialized markets such as the one of 'motor spirits' and finally, after almost a hundred years, fight back coal in the last trenches, electricity production, and steel making." The point, Marchetti told the audience, was to forget about heroic concerns like "the world energy market and world salvation" and to "concentrate on the special case, the special product, the favorable eco-niche." "A new technology," he said, "needs very special conditions to root."

Similar considerations, which still ring true today, led the authors of the massive 1976 Stanford Research Institute study titled The Hydrogen Economy, A Preliminary Technology Assessment to argue that long-range changes affecting entire nations should not be left to private corporations, with their inherent short-term outlook. Corporations formally discount the value of future earnings because their philosophy is essentially piece-meal, thereby pushing the needed ultimate transformation further into the future; that, in turn, requires additional and essentially unnecessary changes at high cost, the authors said: "Decisions in both the private and public sector are generally made with a planning horizon of 5 to 10 years. Corporations formally discount the value of future earnings anticipated from an investment according to the profit or interest the same investment could earn elsewhere. As a result, a dollar that could be earned this year is considered more valuable than a dollar that could be earned next year. Discounting of the future tends to justify a preoccupation with short- rather than long-term goals in decision making."

Because of the massive investments and long lead times required for phasing into a hydrogen economy, "corporations (and governments as well) tend to dismiss hydrogen in favor of concentrating on those activities that continue the viability of the existing order," the authors of the SRI study added. Here they were referring to the attention given during the 1970s to synthetic fuels made from oil shale and coal. Similar short-term strategies were still guiding some major corporations, and the US government, in the late 1990s. The Chrysler Corporation, for example, before its 1998 merger with Daimler-Benz, signed an agreement with the Syntroleum Corporation of Tulsa to develop "designer fuels" manufactured from natural gas that would be inherently sulfur-free and therefore more suitable for fuel cells than sulfur-containing fuels. The US Department of Energy also argues that it would make sense to extract hydrogen from gasoline, and conducts R&D work in support of that objective. The basic rationale is that, in the absence of a hydrogen refueling infrastructure and in view of the technical difficulties of storing hydrogen onboard vehicles, the existing gasoline distribution system should be utilized to provide fuel for fuel cell vehicles—hydrogen—with gasoline functioning as a hydrogen "carrier." Hence sulfur-free "designer fuels." Normal gasoline contains sulfur, which is usually not removed in the refining process. Sulfur doesn't affect the operations of a garden-variety internal-combustion engine, but eventually it destroys the innards of a fuel cell. The obvious advantage would be that such a liquid fuel could be distributed through the existing infrastructure systems. But presumably it would require yet another parallel fuel equipment chain analogous to that for diesel fuel, diminishing the advantage of what at first blush seemed an ingenious way around the infrastructure problem—and it would perpetuate the existing order, the core issue addressed by the 1976 SRI study.

But there were new signs that some of the big US carmakers were losing their appetite for extracting hydrogen from gasoline. At the 1999 Detroit Auto Show, both DaimlerChrysler and Ford said in press releases and in comments by some of their respective executives that extracting hydrogen from gasoline was more difficult than had been anticipated a few years earlier, and that it was an immature technology that would take a long time to get to market. DaimlerChrysler said that reforming methanol into hydrogen seemed a more promising route. However, in still another turn, International Fuel Cells, a subsidiary of United Technologies and Toshiba, said near the end of 1999 that in the previous 18 months it had quietly developed a viable reformer that could recover "98 percent of the energy in the gasoline," and that it had shipped prototypes to carmakers in Japan, the United States, and Europe.[4]

Thus, at present there is no clear preference as to what the fuel should be for the first generation of fuel cell cars.

•

As the 1990s drew to a close, there were signs that the admonitions of hydrogen supporters to look seriously at hydrogen as a clean energy option for the world were beginning to be heeded. Probably the best example is Japan's much-commented-upon WE-NET (World Energy Network) project, which, more than any other national scheme, recognizes the need for long-term planning and for investment in hydrogen energy technology. The

$2 billion WE-NET project—a part of the Global Warming Prevention Plan adopted by Japan's Cabinet in October 1990 after the International Panel on Climate Change targeted CO_2 emission reductions at the 1990 level by the year 2000—was announced in 1993, and was to run for 28 years, through 2020. It is a part of the $11 billion New Sunshine Program, also launched in 1993, which addresses all types of alternative, renewable energies.[5] Annual funding levels have grown steadily in most years, although not always as much and as fast as the program's planners had expected. As reported at the 1996 World Hydrogen Energy Conference in Stuttgart, there were nine major subtasks during the initial 5-year Phase I (which ran from 1993 to the spring of 1999), including water electrolysis for hydrogen production, conversion of gaseous hydrogen into other forms for transport, and fuel cell development. A total of 40 "entities," including universities, think tanks, and corporations, were taking part in Phase I, including five foreign ones. The total funding requested for Phase I was 9.4 billion yen ($78 million at prevailing exchange rates). Phase II, which began in April 1999 and which is to run through Japan's fiscal year 2003, is essentially a continuation of Phase I. The total requested for this segment was 9.7 billion yen ($81 million at prevailing exchange rates). Major projects to be pursued include commercialization of a hydrogen fuel cell power plant, hydrogen fuel storage tank systems for fuel cell vehicles, hydrogen refueling stations based on natural gas reformers, a methanol reforming system, and a PEM electrolyzer.

In November 1997, a month before the Kyoto Global Climate Change Conference, a top-level advisory group to President Clinton told the president that the United States should do more in the areas of renewable energy and hydrogen technology. The 21-member energy panel of the President's Committee of Advisors on Science and Technology (PCAST), the highest-level private-sector science and technology advisory group, had been set up to review the nation's energy priorities, which had not come under scrutiny in many years. It recommended doubling the spending of the Department of Energy's hydrogen program as a part of a general shifting of the DoE's priorities toward greater emphasis on renewables and energy efficiency.

PCAST called hydrogen energy "a very important mid- to long-range research area" and speculated that in the twenty-first century hydrogen might become an energy carrier "of importance comparable to electricity." The panel recommended a $1 billion increase over 5 years in the DoE's annual budget for applied energy technology R&D. It singled out PEM fuel cells as among the most promising of these technologies for "demanding applications such as the automobile."

Finding some fault with the Department of Energy's hydrogen program as it existed then, the PCAST report called for better-articulated near-term, medium-term, and long-term goals. It asserted that continually updated systems analyses assessing alternative energy strategies for the development of a hydrogen economy "could be especially helpful to management in its articulation and periodic updating of these goals."

In his cover letter to President Clinton, John Gibbons, director of the White House Office of Science and Technology and PCAST's co-chairman, said the country needed to improve its R&D effort, especially "in relation to the challenge of responding responsibly and cost-effectively to the risk of global climate change from society's greenhouse gas emissions, in particular, carbon dioxide from combustion of fossil fuels."

In January 1998, the US Department of Energy published a five-year strategic plan for its hydrogen program, which outlined what the program's managers (principally Sig Gronich and Neil Rossmeissl) expected to be the program's goals and hopes over the next 20 years. "Dependence on foreign energy sources is expensive," according to the introduction. "We suffer trade deficits and use our military to protect our energy supply abroad. Environmentally, the Nation is being forced to react to both the need for cleaner urban air and to the potential effects of global climate change. At the same time, while demand growth is projected to slow at home, the growing worldwide demand for electricity will fuel international competitiveness. The solution is a clean, sustainable, domestic energy supply. Hydrogen can be one of the answers." Among its goals, the DoE's plan lists improving the efficiency and lowering the costs of fossil-fuel-based and biomass-based hydrogen-production processes to $6–$8 per million Btu, developing renewable-based, emission-free hydrogen production technologies with a target cost of $10–15 per million Btu, developing fuel cell and reversible fuel cell technologies, and supporting industry in the development and demonstration of hydrogen systems in the utility and transportation sectors.

In Europe, two think tanks, the London-based World Energy Council (WEC) and the International Institute for Applied Systems Analysis (IIASA) in Laxenburg, Austria, collaborated in an effort to get a handle on how the world's energy future might unfold in the twenty-first century. Global Energy Perspectives to 2050 and Beyond, the 1996 study which this effort produced, concluded that the choices about the types of energy that will be used in future decades will be made in the first two decades of the century. As incomes increase around the world, and as the world's population roughly doubles by mid-century, "people will want higher levels of more efficient, cleaner and less obtrusive energy services," said the authors, Arnulf Grübler and Nebojša Nakićenović of IIASA and Michael Jefferson of WEC. Global demand for energy services will grow by as much as an order of magnitude by 2050, but primary energy demands will grow less because of improvements in energy intensities—more primary energy input into a given product or service, replacing physical materials. The report stresses "decarbonization," a term that has increasingly become a part of the international clean-energy vocabulary. "Decarbonization is the single most important fact to emerge from 25 years of energy analysis, and it is now widely accepted," noted one expert, Jesse Ausubel, director of the Program for the Human Environment at Rockefeller University. The term was first coined around 1989 by Kenji Yamaji, a young Japanese researcher who was then working at IIASA; it was introduced into the literature in 1991 by Ausubel, who was cooperating with the IIASA group at the time. Carbon capture, and its twin "carbon sequestration" (meaning safe storage of CO_2 out of atmospheric harm's way) crop up in reports of the International Panel on Climate Change and are gradually finding their way into the media.

Decarbonization is just what it sounds like: taking the carbon out of hydrocarbon fuels. What is left is, of course, hydrogen. Decarbonization will be the industrial end-game strategy of a trend first detected by Cesare Marchetti in the 1970s, when he described a gradual shift, over centuries, from hydrocarbon fuels with high carbon and low hydrogen content (wood, peat, coal) to fuels with increasingly less carbon and more hydrogen (oil, natural gas), culminating, seemingly inevitably, in pure hydrogen as the principal energy carrier of an advanced industrial society. A diagram in an another Nakićenović essay, "Freeing energy from carbon," in the summer 1996 issue of *Daedalus*, showed how the carbon intensity of the world's energy consumption, expressed in tons of carbon per ton of oil equivalent energy, has dropped from just below 1.1 in 1860 to 0.7 in the early 1990s—a decline of about 0.3 percent per year. "The ratio has decreased because high-carbon fuels, such as wood and coal, have been continuously replaced by those with lower carbon content, such as gas, and also in recent decades by nuclear energy from uranium and hydropower, which contain no carbon," Nakićenović said.

"The continued shift to higher quality fuels," Grubler, Nakićenović, and Jefferson wrote, "means continued decarbonization of the energy system. "Decarbonization means lower adverse environmental impacts per unit of energy consumed, independent of any active policies specifically designed to protect the environment." At the global level, "it translates directly to lower CO_2 emissions" but by itself, it is not enough. "Additional active policies will be required."

A shortage of raw fossil energy, a bogeyman and a bane in the days of the oil shocks, is no longer an issue: "The resource scarcity perceived in the 1970s did not occur," said Grübler et al. "With technological and economic development, estimates of the ultimately available energy resource base will continue to increase."

Exactly when will that cleaner, more efficient energy be provided? Theorists looking at planetary energy issues believe it will take a long time.

"The question of what kind of companies will supply energy services, and how, is wide open," according to the WEC-IIASA study. Until about 2020 there isn't going to be much of a change, owing to the long lifetimes of power plants, refineries, and other energy facilities and investments. But during those first 20 years of the twenty-first century, the choices that will determine the physical characteristics of the world's energy economy will likely be made, and those choices will largely determine the course of the following 80 years.

The WEC-IIASA study tried to imagine a wide range of possible pathways to the future. It assumed three basic cases, subdivided into six scenarios (neither predictions nor forecasts, but each an "alternative image of how the future could unfold") that covered a wide range of possible developments. Assumptions ranged from "a tremendous expansion of coal production to strict limits, from a phaseout of nuclear energy to a substantial increase, from carbon emissions that are only one-third of today's level to increases by a factor of more than three."

All the WEC-IIASA scenarios seemed to work and to meet consumers' wants:

For all the variation explored among alternative energy systems, all managed to match the likely, continuing push by consumers for more flexible, more convenient, and cleaner forms of energy. . . . Who their suppliers will be, and which energy sources will be tapped, depends on economic developments in the world, on progress in science and technology, and on policies and institutions. And it depends very much on which suppliers make the near-term decisions that prove more effective in turning the potential match between their supplies and consumer preferences into reality.

All six scenarios foresee a shift toward electricity and to higher-quality fuels, such as natural gas, oil products, methanol, and "ultimately, hydrogen":

No matter what the eventual dominant fuel in a scenario, there is a shift away from noncommercial and mostly unsustainable uses of biomass, and direct uses of coal virtually disappear. Fossil sources continue to provide most of the world's energy well into the next century but to a varying extent across the scenario. Sustainable uses of renewables come to hold a prominent place in all scenarios.

Case A, regarded as fairly unrealistic, envisions a future "designed around ambitiously high rates of economic growth and technological progress," including the conviction "that there are essentially no limits to human technological ingenuity." It also would lead to very high CO_2 levels. Case B describes more modest energy demands, slower technology development resulting in the "most reliance on fossil fuels"—something that "might be described as 'muddling through.'" Case C, the most ambitious one, is broadly "ecologically driven" and is the "most challenging." The future it describes "includes a broad portfolio of environmental control technologies and policies including incentives to encourage energy producers and consumers to utilize energy more efficiently and carefully, 'green' taxes, international environmental and economic agreements, and technology transfer. It reflects substantial resource transfers from industrialized to developing countries, spurring growth in the (global) South. While economic output is less than case A, case C still describes a positive sum game, with total global GDP greater than in case B, and a significant reduction in present economic disparities"—clearly a desirable goal for the planet.

In general, case C postulates that the use of fossil fuels will decline, renewables gradually achieving dominance; that renewables will account for 40 percent of global energy consumption by 2050; and that the percentage will increase to 80 by 2100.

After the mid-century point, the authors state, there are two sharply divergent scenarios under these overall assumptions. One, C2, relies on the development of safe, small scale (150–250 megawatts) nuclear reactors that would be widely accepted, especially in areas of scarce land and dense population, thereby limiting the potential energy supply from renewables. The other scenario, C1, treats nuclear power as a transitional technology that will be virtually phased out by 2100, along with most of the fossil-fuel components.

In general, the authors of the WEC-IIASA study suggest, energy requirements will increase through 2020 roughly in line with population growth (most of it in southern regions) and most of the energy coming from fossil fuels. After that, structural changes in energy systems will become more likely, and some hard policy choices will have to be made. "Uncertainty increases significantly, but not uniformly so," the authors say. "We recognize that no analysis can ever turn an uncertain future into a sure thing," they conclude. "Nor can global scenarios generate customized guidance for individual countries, industries, or firms." Still, they claim to have identified "patterns that are robust across a purposely broad range of alternative future developments."

In another study, titled Decarbonizing the Global Energy System,[6] Grübler and Nakićenović conclude that decarbonization is a continuous, persistent trend, but that it proceeds very slowly, at an average rate of 0.3 percent a year. If the trend holds true, "we might in fact be only half-way through the fossil fuel age that would draw to a close only late in the 22nd century." Globally, the authors said, the energy system is moving in the right direction. But "it will simply not suffice to rely on 'autonomous' structural change toward carbon-freer energy systems. . . . [The energy systems] are dwarfed by historical and anticipated growth rates in energy use and resulting carbon emissions. Substantial acceleration of decarbonization would thus entail both ambitious technological and policy changes."

Cesare Marchetti believes that hydrogen and the emergence of a hydrogen economy have a deeper meaning that goes beyond clean fuel and clean environments. Back in the 1970s, he speculated on the idea that the physical changes associated with making and distributing hydrogen energy would fundamentally affect global politics—an idea that is still fascinating.

Marchetti outlined his ideas at the 1976 Miami Beach hydrogen conference in a 40-minute banquet address titled "From the Primeval Soup to World Government—An Essay on Comparative Evolution." He characterized his startling politico-economic predictions as logical outcomes of the development of a hydrogen economy and as analogous to the physical processes that have shaped the Earth since its beginning. He envisioned large-scale production of hydrogen and a move away from fossil fuels as a planetary liberating process almost comparable to the move away from the primeval soup as energy and life carrier millions of years ago. That move was made possible by the release of tremendous amounts of oxygen through a series of complicated chemical and physical processes. Oxygen formed the ozone layer and, by filtering out lethal ultraviolet radiation, allowed living organisms to come out of the water and conquer the land. "Life had grown to take control of the environment on a global scale," Marchetti asserted. "Now," he continued, "if humanity starts producing hydrogen from water using 'new' sources of free energy—fission, fusion, or perhaps directly tapping the old sun—the moves away from fossil fuels which in my analogy are the equivalent of the primeval soup, what kind of new control of the environment can occur as by-products of this operation?"

In essence, Marchetti foresaw a conflict between the "vertically integrated" power system of nation-states and the "horizontal" power system, covering many states, typical of the multinational corporation. Conflict is inevitable, he believed, but it would not necessarily lead to the destruction of either the states or the corporations. "A state can be seen as a vertically integrated power system filling a geographical area," Marchetti argued.

"A multinational is a horizontal power system organizing a thin layer of human activities without precise geographical boundaries. The horizontal power (the multinational) generates 'confusion' and loss of control by the vertical one (the state), and, with the layers thickening, will inevitably lock with it. The ostensible muck-raking against multinationals is a clear symptom of that. But what will be the outcome?"

Marchetti cited as a historical analogy the relations between the single "multinational" of medieval times, the Catholic Church, and the states of those times. "A fight between geographically bound political powers and a pervasive horizontal power went on in the Middle Ages," he said. The Church was "interfering and competing in many ways with the geographically fragmented political power of the time." Both systems were necessary: "As the two power systems cannot be interchanged, nor either one eliminated, a compromise finally had to be worked out. The political power handed its thin top layer to a supernational power structure, producing a kind of political multinational: the Holy Roman Empire. The maneuvering space of the emperor was politically narrow, but territorially broad and at the proper hierarchical level of abstraction to deal with the Pope on an even basis." "My educated analogy," declared Marchetti, "says that the outcome will be a world government, or more flexibly defined, a world authority to make a dialogue possible." "Energy," he added, "is the largest single business in the world. At 12 dollars per barrel of oil, the turnover is 500 billion dollars a year. It will be 2000 billion dollars at the beginning of the next century. This means that energy multinationals will be the strongest forces in the struggle with political power, and their field of activity a sensitive one. Very large energy centers and energy generation as a world operation are direct consequences of the technological process of water-splitting, and they may modify the political atmosphere, just as oxygen did for the earth atmosphere, leading finally to a world government."[7] He concluded that "the grand design is deploying itself, and we have the privilege and the responsibility of living at a great turning point in history."

As it happened, the global oil business did not grow nearly as much as Marchetti had predicted; in fact, it declined. In late 1998, the total annual dollar volume of the global oil business was reckoned to be in the range of $380–460 billion, assuming production of about 38 billion barrels per day at $10–12 per barrel, according to an oil industry executive. (Of course, by the middle of 2000 oil prices had almost tripled again.)

In "On the patterns of diffusion of innovation," published in the summer 1996 issue of *Daedalus*, Arnulf Grübler asserts that there are basically two innovation strategies. One focuses on incremental changes such as environmental add-on or "end-of-pipe" technologies. These can provide a quick fix, but they "tend to reinforce the dominant trajectory, blocking more systemic and radical changes." The other strategy opts for "more radical departures from existing technologies and practices." These, "such as the development of fuel cells and hydrogen for energy," are "more effective in the long run"; however, they "require much more time to implement because of the multiplicity of forward and backward linkages between technologies, infrastructures, and forms of organization for their production and use." According to Grübler, this interdependence between "individual artifacts and long-lived infrastructure" is at the root of the problem we face:

Within two or three decades, the United States could in principle change its entire fleet to zero-emission vehicles. In fact, 99 percent of vehicles now on the road will be scrapped in this interval. Yet, this interval is too short for the diffusion of the required associated energy supply, transport, and delivery infrastructures, which will inevitably distend the rate of diffusion of end-use devices. Thus, key technologies that we can already envision to raise the quality of the environment probably must await the second half of the twenty-first century to become widespread and influential.

Historically, technology clusters (parallel development of railroads and telegraph, road networks, oil pipelines) have been instrumental in raising productivity and also in alleviating many adverse environmental effects. The emergence of a new cluster could hold the promise of an environmentally more compatible technology trajectory. But it will take time.

"There are," Grübler mused, "times of change and times for change, and unless our individual and collective behavior is modified, these times will remain to frustrate and excite us." This may be so, but it is not only systems analysis and technology and technological advances that will determine our collective future. As the young German diplomat Marcus Bleinroth noted (as quoted in Kempe's *Father/Land*), it is also economics. It is also environmental awareness, still in short supply in some quarters. It is also focused political will—not garden-variety, can-do, pork-and-fundraising political will, but an ethically and humanely informed political will, driven by *enlightened* self-interest, that sees the world as an environmental whole rather than as a series of unconnected boxes and compartments.

There has been a great deal of progress since 1990 in terms of greater environmental awareness and the cleaning up of large parts of our environment. There also has been quite a bit of progress toward the development of clean, environmentally progressive alternative technologies and practices. And we have witnessed the beginning of a race to bring hydrogen-based zero-emission or near-zero emission technologies such as fuel cells to market—a race that both large, long-established manufacturers and small, innovative upstarts have been entering. But in the grand scheme of things all the recent progress seems piddling and feeble when measured against what is needed. In the environment's big picture, Buckminster Fuller's 1969 plaint cited at the start of this book seems more relevant than ever.

Bill McKibben, an environmental writer, expressed these concerns well in a 1999 *New York Times* op-ed piece[8]: "Ten years ago, global warming was a strong hypothesis. Now after a decade of intensive research, scientists around the world have formed an ironclad consensus that we are heating the planet. . . . The question is why we've done so little. In 1992, President George Bush promised the world that the United States would emit no more carbon dioxide in 2000 than it had in 1990. The Clinton Administration instead watched with little apparent concern as our emissions surged more than 10 percent. Congress refuses to even consider the baby step represented by the 1997 Kyoto accords, which would return us to 1990 levels by 2010. The issue barely even crops up in the presidential campaigns."

The reason for the apathy, McKibben believes, "is that we don't yet feel viscerally the wrongness of what we are doing—not just the very rational fears about what it will be like to live in a superheated world but, even more, the simple shock that we've grown so large we can dominate *everything*. . . . Everything that happens above the surface comes at least in part from us, from our appetites and our economies." People will need to be shocked into a "gut understanding" of the environmental situation "if we are to take the giant steps we must take soon." "The sun beating down, the rain pouring down, the wind blowing by," McKibben writes, "are all now human artifacts. We don't live on the planet we were born on. We live on a new, poorer, simpler planet, and we continue to impoverish it with every ounce of oil and pound of coal that we burn."

McKibben's melancholy fin-de-siècle view of the world is, consciously or unconsciously, probably rather widely shared. But there is also an opposite view: a new-century, can-do view, tempered by the insights and experiences of the world's Bill McKibbens, that says "We can and will make the world a better place." This view has inspired and prodded thousands and thousands of activists, researchers, and others to work for a cleaner environment. It was expressed first and best by Jules Verne, in his 1874 novel *The Mysterious Island*:

Yes, my friends, I believe that water will one day be employed as fuel, that hydrogen and oxygen which constitute it, used singly or together, will furnish an inexhaustible source of heat and light. . . . As long as the earth is inhabited, it will supply the wants of its inhabitants, and there will be no want of either light or heat as long as the productions of the vegetable, mineral or animal kingdoms do not fail us.

Notes

Chapter 1

1. R. Buckminster Fuller, *Utopia or Oblivion: The Prospects for Humanity* (Overlook, 1969).

2. United Press International, summarizing Saudi television broadcast of December 21, 1976, after OPEC conference at Doha, Qatar.

3. From transcript of Clinton's remarks at White House Conference on Climate Change October 6, 1997. (In fact Toyota was not working with Ballard Power Systems but was developing its own fuel cells.)

4. Federal Energy Research and Development for the Challenges of the Twenty-First Century, Report of the Energy Research and Development Panel, PCAST, November 5, 1997.

5. "ZEVCO unveils fuel cell taxi, Shell UK chief says company is into hydrogen for real," *Hydrogen & Fuel Cell Letter*, August 1998.

6. During ceremonies at the state capital in Sacramento marking the launch of California Fuel Cell Partnership with DaimlerChrysler, Ford Motor Co., ARCO, Shell, Texaco, California Air Resources Board, and California Energy Commission, April 20, 1999; from press release.

7. "Moving slowly toward energy free of carbon," front page, October 31.

8. The 1998 *Oxford Illustrated Dictionary* defines "pollute" as to "contaminate or defile (the environment)." Hydrogen, when combined (burned, oxidized) with the air's oxygen, produces only water plus minuscule amounts of oxides of nitrogen, inevitable byproducts of any atmospheric burning process. Water doesn't pollute.

9. Others dispute that there is much of a problem. Among academics, one of the most prominent skeptics is Richard Lindzen of the Massachusetts Institute of Technology, who says flatly that "we don't have any evidence that this is a serious problem" (in Global Warming, *New York Times* supplement on Kyoto Climate Change conference, December 1, 1997). Lindzen, described in the *Times* as a "champion to political conservatives and industrial interests who minimize the threat" and "a force of intellectual honesty in a highly politicized debate," told the *Times* that he prizes the environment but that global warming and other issues have

prompted environmental groups to go "off the deep end" and produce "a drum roll that gets rid of perspective." Lindzen's critics, says the *Times*, fault him for professing unwarranted sureness in a field of research rife with uncertainty, and many say he is simply wrong.

10. Global Warming supplement to *New York Times*, December 1, 1997.

11. Hans Deuel, Paul Guthrie, William Moody, Leland Deck, Stephen Lange, Farhan Hameed, Jeremy Castle, and Linda Mearns, "Potential impacts of climate change on air quality and human health," presented at 92nd Annual Meeting of Air and Waste Management Association, St. Louis, 1999.

12. A quad is equal to 8 billion gallons of gasoline—enough to run 16 million cars (more than there are in California) for a year.

13. About 78% of the air we breathe is inert nitrogen; about 21% is oxygen. Air also contains trace gases, including CO_2 (0.03%).

14. For a more detailed explanation of how fuel cells work, see chapter 7.

15. *Encyclopedia of Chemistry*, third edition (Van Nostrand Reinhold, 1973), p. 544.

16. For decades, BMW has been the principal supporter and advocate of this approach.

17. In the context of fuel cells, "reforming" gasoline refers to the technique of extracting hydrogen from the gasoline in a separate reformer before feeding the hydrogen to the fuel cell. Gasoline serves as a hydrogen carrier. In PEM fuel cells, most of the carbon must be removed because it tends to poison that type of fuel cell.

18. However, there is a serious drawback associated with methanol (also known as wood alcohol or methyl alcohol, and used in industry as a solvent, an antifreeze, a denaturant for ethyl alcohol, and a raw material in the synthesis of formaldehyde and other chemicals): it is acutely toxic.

19. For example, the PCAST wants to fund nuclear energy at levels that are multiples of those for hydrogen.

20. Source: personal e-mail communication.

Chapter 2

1. Julius Ruska, in *Das Buch der grossen Chemiker*, ed. G. Bugge (Weinheim: Chemie Verlag, 1974), p. 2.

2. Steven Weinberg, *The First Three Minutes* (Bantam Books, 1977), p. 114.

3. E. Pilgrim, *Entdeckung der Elemente* (Mundus, 1950), p. 144.

4. Nitro-aerial because they occur also during combustion of saltpeter.

5. Richard Koch, in *Das Buch der grossen Chemiker*, ed. Bugge, p. 194.

6. Pilgrim, *Entdeckung der Elemente*, p. 155.

7. Georg Lockemann, in *Das Buch der grossen Chemiker*, ed. G. Bugge, p. 256.

8. Georges Cuvier, in *Great Chemists*, ed. E. Farber (Interscience, 1961).

9. Montgolfières were hot-air balloons named after their inventors, the Montgolfier brothers. They were kept aloft by hot air generated by an open fire in the craft's gondola, not by a buoyant gas such as hydrogen. They were the first balloons ever to rise into the air: In their first flight, launched June 5, 1783, in Annonay, near Lyons, the Montgolfiers' craft stayed aloft for 10 minutes and covered more than a mile. That same year, a montgolfière sailed over Paris in the first manned free balloon flight.

10. J. Pottier and C. Bailleux, "Hydrogen: A gas of the past, present and future," in *Proceedings of Hydrogen Energy Progress VI*, volume 1, ed. T. N. Veziroglu et al. (Pergamon).

Chapter 3

1. *Transactions of the Cambridge Philosophical Society* 1 (1822).

2. My source here is the 1965 Airmont Publishing Company edition.

3. J. B. S. Haldane, *Daedalus or Science and the Future* (Dutton, 1925).

4. The latter was a reference to the first fuel cell, constructed in 1839 by the English philosopher-physicist-lawyer William Grove, which produced water from oxygen and hydrogen but almost no electricity. Development of modern fuel cells that also produce electricity did not start until the 1950s.

5. The Fischer-Tropsch process synthesizes various hydrocarbons from carbon monoxide and hydrogen. The Bergius process, which breaks down coal into a synthetic crude oil with the help of hydrogen and a catalyst, enabled Nazi Germany to produce large amounts of synthetic aviation gasoline.

6. There was a happy footnote, however: In 1978, Erren showed up at the World Hydrogen Energy Conference in Zurich. When he was introduced at the final session, the applause was loud and long—a fitting tribute for a man who had been 40 years ahead of everybody else.

7. One exception was R. O. King, who continued his research on hydrogen combustion at the University of Toronto in the late 1940s and the 1950s.

8. Marchetti's interest in hydrogen started in a roundabout fashion back in the late 1950s, when he was working for the Italian national oil company, Ente Nazionale Idrocarburi (ENI), trying to find ways to make use of the off-peak power of nuclear reactors. "The idea then was to somehow use electricity to make adenosine triphosphate (ATP) which is the energy carrier of biological systems," he recalled in an early interview. Marchetti tried to find ways to make synthetic ATP to act as intermediary between an electrical energy source, like a reactor, and a biological system. Eventually Marchetti found "this didn't work because ATP never leaves the individual cell, therefore you cannot put it into a cell." In the 1960s, he recalled that in his high school days he had read that there were certain bacteria able to metabolize hydrogen and oxygen and grow on that: "So I thought this material would be the right interface between food chain and electricity because you can make hydrogen

and oxygen electrolytically." A bibliographic search of the literature under H—for hydrogen—turned up all kinds of chemical processes "that could do almost everything—reduce materials in chemistry, reduce minerals to metals, run engines and things like that." Looking at the problem more generally, "I decided from a systems point of view electrolysis was not very good for a number of reasons. So, going to the root of the problem, that of transferring free energy from a heat source to a chemical system, I decided that the best way was to go to a chemical system via a multi-step process." That was about 1969, and the way he tells it with a smile, "I told De Beni to invent such a system, and he did invent one."

9. Linden was one of the first energy experts to use the term "hydrogen economy." See *Journal of Fuel Heat Technology* 18 (1971), November, p. 17.

10. I was working at the time in Milan, Italy, as bureau chief for McGraw-Hill World News, the company's in-house news service for most McGraw-Hill technical and business magazines, including *Business Week*. I had suggested the piece after interviewing Marchetti at the Euratom Research Center in Ispra, some 40 miles up the road from Milan in the foothills of the Italian Alps. That taped interview in which Marchetti laid out the basic concept of a hydrogen economy marked the beginning of my interest in hydrogen.

11. In a 1976 interview, Escher recalled: "I became very much interested in hydrogen because it was paying my salary in 1958 when I began work on hydrogen-oxygen rockets at NASA Lewis [Research Center]. My job was to help run tests and lay out thrust chambers that worked on hydrogen and oxygen which was considered a far-out rocket fuel." Escher's "transition to real life," as he called it, came in the late 1960s, when he went to work for Rocketdyne, a maker of rocket equipment in California, and was assigned to sell rocket technology to the outside world. He recalled: "I made a few trips to the public utilities and other such people with potential hydrogen interests. It was interesting, and immediately it churned up into a big activity, and a year later, in 1970 I left and became an independent consultant (on hydrogen and hydrogen-related alternative energy systems)." At the end of the 1990s, after several career moves, Escher joined the Science Applications International Corporation in Huntsville.

12. Bockris, Escher, Marchetti, and Weil have already been introduced in this volume. Tokio Ohta taught in the Department of Electrical Engineering at Yokohama National University in Japan; Van Vorst lectured at the School of Engineering and Applied Science at the University of California at Los Angeles; Martinez worked for Venezuela's national oil organization in Caracas; Seifritz was a lecturer at the Swiss Federal Institute for Reactor Research; Abdel-Aal was with the College of Petroleum and Minerals in Dharan, Saudi Arabia.

13. The DoE, established in 1977, combined several other agencies, including the Energy Research and Development Administration, set up in 1974.

14. Project Sunshine's other main research areas were solar energy, geothermal energy, coal gasification and liquefaction.

15. Senators Spark Matsunaga (D-Hawaii) and Daniel Evans (R-Washington) were the Senate's most prominent hydrogen advocates at the time.

16. A real foulup occurred in 1986, when $1 million earmarked for the hydrogen program at Brookhaven National Laboratory simply vanished in the Congressional budgeting process. The word from Capitol Hill was that this was due simply to an editing error, not because anybody wanted to do in the hydrogen program. "It fell between the cracks," said a Capitol Hill source at the time. Ultimately, the missing million was restored. After 1987, it took 10 years to restore the DoE's hydrogen program to a semblance of reasonable funding, with $15 million appropriated in both 1996 and 1997—still small potatoes compared to other renewables such as photovoltaics which received close to $60 million in fiscal 1997. (In fairness, it should be mentioned that hydrogen-related areas such as fuel cells got in the neighborhood of $70 million in fiscal year 1998—about one third for transportation and the rest for stationary fuel cells.)

17. Others have estimated higher costs. In a December 1997 article in the *Atlantic Monthly*, Paul Ehrlich and four co-authors say the price of gasoline carries "a social cost of *at least* $4.00 a gallon but is sold to Americans for $1.20." Senator Richard Lugar (R-Indiana), chairman of the Senate Agriculture, Nutrition and Forestry Committee, estimated the real cost of Middle Eastern crude oil to be at least 4 times the going price. "The effects of world dependence on Middle Eastern oil means that while the quoted market price per barrel is about $20, the cost associated with keeping shipping lanes open, rogue states in check and terrorists at bay, may more than quadruple the price per barrel," Lugar said in his opening statement at a November 13, 1997 committee hearing on energy security and global warming.

Chapter 4

1. Zweig cites data published in the mid 1990s by Jane Hall of Fullerton State College and David Abbey of the Loma Linda Medical Center School of Public Health showing that the societal costs of these illnesses was about $14 billion per year in the Los Angeles Basin alone.

2. Over 10 years, the partners invested about 145 million marks ($89 million) in the facility.

3. Institute of Gas Technology, Survey of Hydrogen Production and Utilization Methods, 1975.

4. The energy content of fuel gases is typically expressed as higher (gross) heating value (HHV) or lower (net) heating value (LHV). Both measure the heat—number of British thermal units (Btu)—produced per standard cubic foot (scf) of the gas in the complete combustion of the gas, at constant pressure, and with the product of the combustion measured at 600°F and all water formed by that combustion condensed into liquid again. The lower heating value uses the same data, but the water does not condense again into a liquid but stays as a vapor. Normally, the difference is not significant for most fuels used in conventional combustion systems; for hydrogen, however, the difference is about 15.6%. This is important since fuel cells can capture and utilize only the LHV of hydrogen.

5. SPE is a a registered trademark owned by Hamilton Standard, a division of United Technologies Corp. Hamilton Standard was renamed Hamilton Sundstrand after United Technologies acquired the Sundstrand Corporation in 1999.

6. J. Funk and R. Reinstrom, "Energy requirements in the production of hydrogen from water," *I&EC Process Design and Development* 5 (1966): 336–342.

7. Of that, the United States produces about 10 million tons—most of it as a "captive" commodity in a refinery or a chemical plant making ammonia, a basic component of fertilizer, where it is used without ever leaving the plant.

8. Epyx (a spinoff of Arthur D. Little, Inc.) became Nuvera in early 2000 in a merger with the Italian firm De Nora.

9. The US government's much-maligned Syn-Fuel program of the 1970s, which attempted to make synthetic gasoline from fossil sources other than crude oil and which was terminated in 1986, may conceivably see the light of day again in some new and improved version.

10. Robert Williams, Eric Larson, Ryan Katofsky, and Jeff Chen, "Methanol and Hydrogen from Biomass for Transportation, with Comparisons to Methanol and Hydrogen from Natural Gas and Coal," report 292, Center for Energy and Environmental Studies, School of Engineering and Applied Science, Princeton University, 1995.

11. The Clinton administration has already taken first steps encouraging cultivation of energy crops: Up to 500 farmers and landowners participated in a 4000-acre pilot demonstration program to grow switchgrass for power production announced by the Department of Energy in October 1996. Energy was to be produced either via a 6-MW gasifier or as additional fuel in a mundane pulverized-coal boiler.

12. C. Thomas, Brian James, Frank Lomax, Jr., and Ira Kuhn, Jr., "Fuel options for the fuel cell vehicle: hydrogen, methanol or gasoline?" *International Journal of Hydrogen Energy* 25 (2000): 551–567.

13. Graetzel's initial project was funded by the Swiss Energy Ministry, by Sandoz, and by Asea Brown Boveri.

14. Graetzel is also collaborating with scientists at the National Renewable Energy Laboratory in Golden, Colorado.

Chapter 5

1. With some obvious political and economic advantages—were hydrogen available in many parts of the world, oil cartels such as the OPEC of the 1970s would have a difficult time enforcing their will.

2. At the DoE's 10-MW Solar II solar tower in Barstow, California, solar heat is stored at about 1050°F (573°C) very efficiently in tanks containing molten salt. About 98% is the annual average round-trip efficiency. This assumes the plant runs every day dispatching energy to the grid, using solar energy collected during the day at night stored as a heat in large (3 million pounds, 40 feet high, 25 feet in diameter)

tanks, says James Pacheco, principal member of the technical staff at Sandia National Laboratories, which operates Solar II together with the National Renewable Energy Laboratory. Pacheco says heat dissipates only very slowly: The tank holds about 105 MWh of electricity, and when not in use it loses only about 100 kw per hour, or about 5°F per day. "It's a very simple design, and very cheap," says Pacheco.

3. Since retiring in the 1990s, Bockris has shifted his attention to more esoteric fields, including low-temperature nuclear reactions, quantum electrochemistry, and transmutation of metals.

4. J. Bockris, *Energy: The Solar-Hydrogen Alternative* (Halstead, 1975).

5. Although no new reactors are being built in the US, and many were canceled in the 1970s and the 1980s, the 104 reactors that were operating in the US in 1999 were operating more efficiently (with capacity factors approaching 80% in the 1990s compared to just above 50% in the 1970s) and were generating more of the nation's electricity (almost 25%, compared to about 10% in the 1970s) than ever before, according to the *New York Times*. A March 7, 1999 *Times* article on the twentieth anniversary of the Three Mile Island accident said that "the nuclear industry will be around for years to come, but seems to have peaked in terms of the number of working reactors and their share of power generated nationally." The last reactors (Palo Verde 1, 2, and 3, in Arizona) were ordered in October 1973, the month the Arab oil embargo began, the article said. Worldwide, 437 reactors were operating at the end of 1997, and 36 were under construction.

6. In *Renewable Energy—Experts and Advocates*, ed. T. Dickerman (American Association for Fuel Cells, 1997), a resource book for American high school debaters.

7. I am indebted to C. E. (Sandy) Thomas, vice president for energy and environment of Directed Technologies, Inc., for this succinct summary of fusion's potential. Sandy worked for 8 years in fusion research, then for 5 years investigating amorphous silicon photovoltaics, before going to Washington, where he spent another 8 years as energy aide to Sen. Tom Harkin. He moved to Directed Technologies in the late 1990s.

8. Ogden says that as a rough measure, 1 million scf per day fuels about 800 PEM-fuel-cell cars each day, or about 80 PEM-fuel-cell buses.

9. Linda Church Ciocci, "Hydropower, The Nation's Leading Renewable Energy," in *Renewable Energy—Experts and Advocates*, ed. Dickerman.

10. Source: "Iceland and Daimler-Benz/Ballard start plans for hydrogen economy," *Hydrogen & Fuel Cell Letter*, June 1998.

11. A gigawatt is 1000 megawatts; a terawatt is 1000 gigawatts.

12. Email, August 22, 2000.

13. Randall Swisher, "Wind Energy: Technology, Resource Potential and Policy Needs," in *Renewable Energy-Experts and Advocates*, ed. Dickerman.

14. *Solar Energy—Today's Technologies for a Sustainable Future*, ed. M. McIntyre (American Solar Energy Society, 1997).

15. Source: David Kearney of Kearney & Associates, a former Luz vice president.

16. Source: James Rannels, acting director of Photovoltaic and Wind Energy Division of US Department of Energy, in *Renewable Energy—Experts and Advocates*, ed. Dickerman, p. 51.

17. Ibid.

18. Donald Osborn, photovoltaic section, *Solar Energy*, ed. McIntyre.

19. Source: Ralph Overend and Susan Moon, "Biomass," in *Solar Energy*, ed. McIntyre.

20. Ibid.

21. Ibid.

22. Ibid.

23. "Weizmann Institute Scientists Zero In on Direct High-Temperature Solar Water Splitting," *Hydrogen & Fuel Cell Letter*, November 1996.

24. Jörg Schlaich, *The Solar Chimney—Electricity from the Sun* (Edition Axel Menges, 1995).

25. Source: Wolfgang Schiel, a physicist working in Schlaich's office.

26. As of mid 1999, the project was on hold, possibly because of financing uncertainties which may have had something to do with skittishness in financial markets due to the recent nuclear tests in India and Pakistan.

27. Source: "Energie aus dem Weltall mit der 'Kraftsoletta,'" *Handelsblatt*, September 6, 1977.

28. There are two basic types: "closed-cycle" and "open-cycle." In a closed-cycle system, warm surface seawater and cold deep seawater are used to vaporize and condense the working fluid such as ammonia, which then drives the turbine generator in a closed loop. In an open-cycle system, surface seawater is flash-evaporated in a vacuum chamber, and the resulting low-pressure steam drives a turbine-generator. Cold seawater is then used to condense the steam after it has passed through the turbine. The open cycle, therefore, can be configured to produce fresh water as well as electricity.

29. Vega does not consider mere paper studies "as activity, given the numerous studies, in at least seven different languages, in the last century." Source: Fax from Luis Vega, November 12, 1997.

30. "Strategic Plan for the Geothermal Energy Program," June 1998.

31. There are two main types of geothermal power plants: flash steam type and binary cycle. The flash steam plant is driven by pressurized, hot (300–700°F) steam brought up from depth of as much as more than 10,000 feet. When this pressure is reduced at the surface by about a third or so, the water "flashes"—explosively boils—into steam which drives the turbine and generator. Binary plants are driven

by lower-temperature (212–300°F) geothermal fluids. Here, the fluids are passed through a heat exchanger which heats a secondary, usually organic, working fluid such as isopentane which turns into steam at temperatures lower than water. That organic steam drives the turbine and the generator before being recondensed into fluid and vaporized again by geothermal heat.

Chapter 6

1. *Los Angeles Times*, January 9, 1997.

2. "Can Motor City come up with a clean machine?" *New York Times*, May 19, 1999.

3. US Energy Information Administration, Annual Energy Review, 1997 (1999).

4. US Environmental Protection Agency, National Air Quality and Emissions Trends Report (1997).

5. Office of Integrated Analysis and Forecasting, Emissions of Carbon from Energy Sources in the US—1998 Flash Estimate, June 1999 (www.eia.doe.gov/oiaf/1605/flash/flash.html).

6. Buses and centrally fueled fleets of trucks or vans are a different matter; hydrogen may well be the fuel of choice for them.

7. Source: personal communication.

8. There are other types, such as chemical hydrides and liquid organic hydrides.

9. K. Hoffman, W. Winsche, R. Wiswall, J. Reilly, T. Sheehan, and C. Waide, Metal Hydrides as a Source of Fuel for Vehicular Propulsion, technical paper 690232, Society of Automotive Engineers, 1969.

10. Personal communication.

11. "High oil prices, Iceland plans, skepticism about small home PEMs mark annual NHA meeting," *Hydrogen & Fuel Cell Letter*, April 2000.

12. E. Dickson, T. Logothetti, J. Ryan, and L. Weisbecker, "The Use of Personal Vehicles Within the Hydrogen Energy Economy—An Assessment," in proceedings of Hydrogen Economy Miami (THEME) Conference, 1974.

13. E-mail communication from Volkswagen AG archives.

14. US Department of Health and Human Services, Public Health Service, Agency for Toxic Substances and Disease Registry, *Methanol Toxicity*, TSDR Case Studies in Environmental Medicine, April 1997.

15. Fritz Kalhammer, Paul Prokopius, Vernon Roan, and Gerald Voecks, Status and Prospects of Fuel Cells as Automobile Engines—A Report of the Fuel Cell Technical Advisory Panel, prepared for State of California Air Resources Board, 1998.

16. "CARB Fuel Cell Study Is 'Unduly Pessimistic' about Hydrogen for Cars, Conference Is Told," *Hydrogen & Fuel Cell Letter*, March 1999.

17. "Selling Fuel Cells," *The Economist*, May 25, 1996.

18. The April 1999 launch of the California Fuel Cell Partnership was a major milestone. Headquartered in Sacramento, the Partnership brought together eight carmakers (DaimlerChrysler, Ford, General Motors, Honda, Hyundai, Nissan, Toyota and Volkswagen); federal and state government agencies (the US Departments of Energy and Transportation; the California Air Resources Board, the California Energy Commission, and the South Coast Air Quality Management District); fuel cell makers (Ballard Power Systems and International Fuel Cells); energy companies (BP, Shell, and Texaco); hydrogen suppliers (Air Products and Chemicals, Inc., Germany's Linde AG, and Praxair); methanol fuel supplier Methanex; and two California transit agencies (the Alameda-Contra Costa Transit District in the San Francisco Bay area and SunLine Transit in the Palm Springs area).

19. Romano helped manage and direct NASA's Lunar Roving Vehicle Project in the 1960s.

20. Other researchers say that higher temperatures are needed: at least 260°C for methanol, and 700–1000°C for other fuels, depending on the type of catalyst used.

21. Today, PEM technology is the clear favorite for such uses. With the exception of the Toyota effort and the Renault-Peugeot effort, most of the efforts depend in one way or another—so far at least—on Ballard technology. Daimler-Benz, Honda, Nissan, Volkswagen, and Volvo have had contracts with Ballard at various times.

22. The Bavarian government, which is trying hard to encourage hydrogen technology as a high-payoff future technology within its borders, assigned coordinating responsibility for the project to Ludwig-Boelkow-Systemtechnik, a subsidiary of the Ludwig-Boelkow Foundation near Munich, which has been active in promoting and supporting hydrogen energy technologies since the early 1980s.

23. Gasoline-powered golf carts were banned in California in 1997.

24. The problems were symptomatic of the fears and misconceptions that continue to dog hydrogen energy technology. Lehman cited an incident that had occurred during a civic meeting in Palm Desert: after a speaker had shown a slide of some fifth graders looking at a picture of the proposed solar-powered hydrogen facility, a visibly distraught audience member had asked "How can you even think about having children involved in solar hydrogen in light of the Hindenburg accident?"

25. James Cannon, *Harnessing Hydrogen—The Key to Sustainable Transportation* (INFORM, 1995), p. 26. There have also been at least a few hydrogen-powered motorcycles and jet engines.

26. Christoph Huss and Ragna Nordheimer, "The BMW Hydrogen Research Vehicle—A Forerunner of the Hydrogen Society," XII World Hydrogen Energy Conference, 1998, Buenos Aires.

27. In contrast, and illustrative of the wide divergence of opinions among experts and of the rapid pace of change, a paper presented at a London fuel cell conference in late 1999 claimed that DaimlerChrysler estimated the cost of the fuel-cell system in NECAR 4 to be about $545 ("Economics, fuels, catalysts highlight 'F-99' conference agenda," *Hydrogen & Fuel Cell Letter*, November 1999).

28. Cryogenic hydrogen-storage technology offers performance roughly comparable to that of conventional gasoline engines, according to Huss: BMW's latest generation of double-walled cryogenic tanks, with 70 layers of aluminum film sandwiched between glass-fiber mats, weigh only about 60 kilograms when filled with 140 liters of LH_2, giving a range of 250 miles.

29. "Alternative Energy Sources for Road Transport—Hydrogen Drive Test," published by Verlag TUV Rheinland for Federal Ministry for Research and Technology.

30. In the more sophisticated "internal mixture formation," hydrogen is injected directly into the combustion chamber during combustion, as in a diesel engine. For this, hydrogen must be supplied at high pressure. This approach eliminates backfire and self-ignition in the compression phase, and there is no loss of volumetric efficiency. Engine operation is characterized by high power density and low fuel consumption, according to the TUV report, but the system is technically more complicated and problems may occur with mixture homogenizing at higher engine speeds. It is the approach favored by BMW, which right from the beginning of its work on hydrogen i.c. engines has concentrated on the use of liquid hydrogen.

31. Source: "Alternative Energy Sources for Road Transport—Hydrogen Drive Test" (TUV report).

32. DFVLR was later renamed DLR.

33. Peschka and BMW parted company shortly after the completion of their the first vehicle.

34. The Perris Smogless Automobile Association ceased its work in 1973. Paul Dieges maintained his involvement with hydrogen through the late 1990s, working with the Southern California chapter of the American Hydrogen Association.

Chapter 7

1. A. Appleby and F. Foulkes, *Fuel Cell Handbook* (Van Nostrand Reinhold, 1989; Krieger, 1993). This book, one of the standard reference texts on fuel-cell technology, provided much of the historical and technical information for this section. Another main source, especially for the early history of fuel-cell technology, is *Fuel Cell Systems*, ed. L. Blomen and M. Mugerwa (Plenum, 1993).

2. Karl Kordesch and Guenter Simader, *Fuel Cells and Their Applications* (VCH, 1996).

3. Hydrazine, ammonia, and methanol have been investigated as "carriers" of hydrogen for use in fuel cells, in part because hydrogen is difficult to store onboard. Of the three, hydrazine, a liquid, was found to be the most promising for fuel-cell use because it is easily dissolved in aqueous electrolytes—the electrochemical "heart" of the fuel cell—and because it is readily oxidized. However, its potential use for normal ground transportation is unlikely because it is expensive to manufacture: Appleby says the cost is about 15–20 times as high as for hydrogen for the same amount of energy. But even if cost were not a factor, its suitability for normal

road transport with its everyday hazards ended when it was found that the stuff is highly toxic—human tolerance is about 1 ppm in air—and that it poses a severe explosion hazard when exposed to heat or when it reacts with oxidizing materials. Methanol, despite some biomedical and environmental hazards, is currently the fuel of choice. Ammonia, the least reactive of the three, also has good points in its favor. It is available worldwide (as fertilizer, for instance), costs are only slightly higher than those of methanol on a Btu basis, and it is easy to handle under low pressure. It has low toxicity, it is easy to handle as a pure fuel in lightweight containers—and leaks would be easily detected because of its pungent smell. It hasn't received much attention as a power source for fuel cells except for remote or specialized application, but some speculate this may change with future vehicle applications—an ammonia-fueled tractor for farmers?

4. Even a cursory description of the many programs that got underway in those years would require much too much space here; interested readers should see the books by Appleby and Foulkes, Blomen and Mugerwa, and Kordesch and Simader for details.

5. Tokyo Electric Power Company, Inc. (Principal Investigator K. Shibata), Demonstration Testing of 11 MW Phosphoric Acid Fuel Cell Power Plant—from Planning to Power Generation. Interim Report for Electric Power Research Institute, Palo Alto, 1992.

6. Source: email from Akifusa Hagiwara, Manager, Material Science Group, Energy and Environment R&D Center, Tokyo Electric Power Co., October 12, 1999.

7. Westinghouse's "tubular solid oxide" technology, generally regarded as the most advanced of this type, was acquired by Siemens in 1998 as part of Siemens's purchase of Westinghouse's power-generation subsidiary. Siemens subsequently stopped work on its own version, which was called "planar SOFC technology."

8. The American Power Corporation was a commercial offshoot of Analytic Power, which has been developing fuel-cell systems for the US military and for civilian clients.

9. Christopher Dyer, "Replacing the battery in portable electronics," *Scientific American*, July 1999, p. 88. Dyer's article was one of three in a special report on fuel cells in that issue. The second (A. John Appleby, "The Electrochemical Engine for Vehicles") described the prospects of fuel cells as a power source for vehicles. The third (Alan Lloyd, "The Power Plant in Your Basement") described the outlook for small fuel cells powering homes and other small-scale applications.

10. "Bell Researcher Develops Thin-Film Fuel Cell," *Hydrogen Letter*, February 1990.

11. Because they are very quiet, and because they are inherently more efficient than diesel or other i.c. power plants, fuel cells have long been regarded by the navies of the United States, Germany, Canada, Australia, and Sweden as an attractive means of propelling small submersibles (such as unmanned underwater vehicles used for marine research, pipeline repair, salvage, and exploration), and even full-size submarines.

12. Additionally, Kordesch and Simader differentiate among different fuel-cell systems: direct fuel cells (those that electrochemically convert fuels such as hydrogen, but also organic or nitrogenous compounds [ammonia, hydrazine], even metals and hydrogen/halogen combinations at low, intermediate or high temperatures); indirect fuel cells (the type that requires a reformer to extract hydrogen, or a fuel cell running on biochemical fuels that are decomposed via enzymes to produce hydrogen); and regenerative fuel cells (systems that can work either as a fuel cell producing electricity, or as a reversible system—as an electrolyzer—in which electrical energy can be stored by splitting water, for example, into hydrogen and oxygen).

13. Also known as polymer electrolyte fuel cell (PEFC). Another term occasionally used is ion exchange membrane (IEM).

Chapter 8

1. The standard TU-134, still in service today in China and elsewhere, weighs 188,000 pounds, carries a crew of three, has a maximum speed of 584 mph, and can carry 164 passengers.

2. Years later it emerged that the Russians had in fact developed a system for pumping liquid hydrogen. Daimler-Benz Aerospace, which in the mid 1990s embarked on a project of demonstrating a liquid-hydrogen-powered commuter plane, said in a brochure that the new plane's tank pump would be based on the TU-155's pump, but that it would be redesigned because the old Russian pump had produced too much heat, resulting in too much boiloff of fuel.

3. The engine of Conrad's plane was essentially standard, though its fuel injection system had been modified (with help from G. Daniel Brewer) to handle liquid hydrogen. The timing and control systems for actuating the injection valves were devised by Roy Parsons, an electronics design engineer from Pompano Beach. The tank had been fabricated and donated by Consolidated Precision Corporation of Riviera Beach. The flight was preceded by several months of ground testing.

4. Parts of the hydrogen fuel system from Conrad's plane now rest in Phoenix, in a small warehouse, maintained by the American Hydrogen Association, that has been turned into a museum of early hydrogen economy artifacts.

5. A paper by Brewer describing the various designs was published in the first edition of the *International Journal of Hydrogen Energy* in early 1976.

6. The reasons for the difference in tank arrangements have to do with the fuels' characteristics. JP-4 is stable at ambient temperatures, and the shape of the tank therefore doesn't make much difference, permitting JP-4 to be stored in the wings, which also helps to stabilize and strengthen the wings structure. Supercold liquid hydrogen, on the other hand, boils off, creating pressure gas. To contain this pressure and to minimize the "chinks in the armor"—odd angles and curves where pressure might build to the bursting point—a round, or nearly round, perhaps elliptical, pressure tank is best.

7. This works for both hydrocarbon fuels and hydrogen, but hydrogen is especially suitable because it mixes more easily and therefore a simpler combustor design could be utilized.

8. A good source of information on these early efforts is Russell Hannigan's *Spaceflight in the Era of Aero-Space Planes* (Krieger, 1994).

9. At the 1996 World Hydrogen Energy Conference, held in Stuttgart, a paper from Iwatani International Corporation, one of Japan's leading industrial gas manufacturers and distributors, said that continuous mass production by this method was possible but that further detailed studies on storage, flowability, thermal conductivity, and commercialization technologies were still required.

10. It is not likely that any NASP-type transatmospheric craft was ever built. However, the possible secret existence of a very fast (Mach 4.5–6), very high-flying aircraft powered by liquid hydrogen, liquid methane, or some other unconventional fuel has excited the imagination of enthusiasts and professional analysts. There has been speculation that this plane, generally referred to as "Aurora," is a successor to the supersonic SR-71 reconnaissance plane. The US government has steadfastly denied its existence. The Federation of American Scientists' aerospace expert, John Pike, devotes some 16 pages to "Aurora/Senior Citizen" on the "Mystery Aircraft" section of the FAS Intelligence Resource Program web site, where he cautiously notes that "although there is a growing body of evidence that could be interpreted to suggest the existence of one or more advanced aircraft behind the veil of government secrecy, the evidence remains suggestive rather than conclusive." There have not been any reported sightings or other developments for several years, Pike noted in a fall 1999 email message, adding that "the trail has gone cold."

11. *International Journal of Hydrogen Energy* 4 (1979), no. 3.

12. As the Bonn correspondent for McGraw-Hill World News and for *Business Week* at the time, I reported the story.

13. "DaimlerChrysler Aerospace Launches Europe-Wide LH2 Airplane Systems Study," *Hydrogen & Fuel Cell Letter*, August 2000.

14. According to a report issued in August 2000 by Britain's Institute for Public Policy Research (Chris Hewett and Julie Foley, Plane Trading: Policies for Reducing the Climate Change Effects of International Aviation), "aviation is the fastest growing source of transportation greenhouse gases, although it is still small in proportion to others."

15. "NASA Launches Hydrogen Airplanes Study, Includes Fuel Cells for Propfan Planes," *Hydrogen & Fuel Cell Letter*, September 2000.

16. "Fast Forward" column, *Hydrogen & Fuel Cell Letter*, September 2000.

Chapter 9

1. Pyle has collected these plus a few related articles from *Home Power* in a 126-page booklet titled *Solar Hydrogen Chronicles—A Hands-On Guide to Solar*

Hydrogen Fuel. It is available from his company, H-Ion Solar Inc. (6095 Monterey Avenue, Richmond, CA 94805; phone (510) 237-7877).

2. "Swiss architect converts home to hydrogen," *Hydrogen Letter*, July 1990.

3. According to a report given at the 1996 World Hydrogen Energy Conference, Fraunhofer Institute scientists working on this project also developed several types of hydrogen burners with very low NOx emissions, including a diffusion-type burner for retrofitting conventional natural gas-type burners. They also fabricated and tested a 33-kW gas absorption chiller and a 21-kW space heater.

4. "Five years in the making, $18 million hydrogen production/fueling station opens in Munich," *Hydrogen & Fuel Cell Letter*, June 1999.

5. "Three SOFC power plants under way in Europe and US, guessing game about fourth 1 MW unit," *Hydrogen & Fuel Cell Letter*, August 1999.

6. By one count, some 60 companies—DaimlerChrysler's chief of fuel cell development, Ferdinand Panik, cited that figure in his presentation at the company's "Innovation Symposium" in November 1999 in Stuttgart ("DaimlerChrysler to Roll Out Next-Generation Necar 5, Shows Miniaturized Stack, Reformer," *Hydrogen & Fuel Cell Letter*, December 1999).

7. "Unable to raise new R&D funding, MC Power is forced to shut down," *Hydrogen & Fuel Cell Letter*, March 2000.

8. Indirect proof that very light "leaky" gases can be success fully stored in underground formations has come from several years of work with helium by the US Department of the Interior. Helium has been injected into depleted natural gas fields near Amarillo. Hydrogen advocates believed the data from these trials show that hydrogen could also be stored in much the same way.

9. "Team develops hydride car A/C system," *Hydrogen Letter*, June 1988.

10. "Scientists seek funds for hydride bus AC," *Hydrogen Letter*, May 1990.

11. "Savannah River, Thermal Electric Devices sign CRADA for hydride refrigeration system," *Hydrogen & Fuel Cell Letter*, December 1995.

12. G. Kaske, P. Schmidt, and K.-W. Kannengiesser, "Vergleich zwischen Hochspannungsgleichstromübertragung und Wasserstofftransport" (Comparison between high-voltage DC transmission and hydrogen transport), VDI Energy Technology Society Meeting, Stuttgart, 1989.

13. Joan Ogden, *Renewable Hydrogen Energy System Studies* (Center for Energy and Environmental Studies, Princeton, 1993).

Chapter 10

1. Robert Hazen, *The Alchemists: Breaking Through the Barriers of High Pressure* (Times Books, 1993).

2. "Big gun makes hydrogen into a metal," *New York Times*, March 26, 1996. A recent, more detailed account by Nellis is the May 2000 cover story of *Scientific*

American, "Metallic Hydrogen—The Stuff of Jupiter's Core Might Fuel Fusion Reactors."

3. "The thrust produced per unit rate of consumption of the propellant usu. specified in pounds of thrust per pound of propellant used per second and forming a measure of the efficiency of performance of a rocket engine." (*Webster's Third New International Dictionary*)

4. According to William Fife, Lavoisier—one of the discoverers of hydrogen—discussed the possible use of hydrox in one of his seminal studies, published by France's Academy of Sciences in 1789.

5. *Knallgas* is German for detonating gas, specifically a mixture of hydrogen and oxygen. It is sometimes used in the English-language technical literature.

6. It is absorbed by some animals, though. To remove it for human consumption, the bacterial mass is heated quickly to 60°C, activating certain enzymes that break up RNA. These components can be removed from the material by centrifugation.

7. Alexander Steinbüchel and Bernd Füchtenbusch, "Bacterial and other biological systems for polyester production," *TiBtech* 16, October 1988.

8. The definition of the "water hole" is actually a bit wider. The The Oak Ridge Observatory in Harvard, Massachusetts tunes in to frequencies in the range from 1400 to 1720 MHz, a spectrum that takes the Doppler effect into account.

Chapter 11

1. Brewer based his conclusions on the report by the Spanish Ministry of Transport and Communications which said, among other things, that the KLM plane's fuselage was not particularly deformed, and neither the impact against the Pan Am plane nor the one against the ground were particularly violent. Brewer thought that under these circumstances any hypothetical liquid-hydrogen tanks, which in some Lockheed designs of those years would have been located inside the fuselage in two tanks, one forward and one aft of the passenger compartment rather than in the wings as in standard, commercial kerosene-fueled B-747s, would probably not have been ruptured—or at least not both of them. The only fire would have been due to the relatively small amounts of hydrogen in the feedlines running through the wings to the engines.

2. However, even a weak spark, such as one caused by static electricity from a human body, is enough to ignite any of these fuels. Such a spark produces about 10,000 microjoules.

3. Source of data: Jesse Hord, Is Hydrogen Safe? (technical note 640, National Bureau of Standards, 1976).

4. Westinghouse Savannah River has long had experience with hydrogen and hydrides as a maker of nuclear explosives, and in recent years the company has become active in developing peacetime applications such as a hydrogen bus and the use of hydrides in air conditioning equipment.

Chapter 12

1. Other institutions, including Iceland's National Power Company, Landsvirkjun, were expected to join later.

2. For example, in 1976 Marchetti proposed Energy Islands, truly gargantuan nuclear energy facilities hidden away in remote parts of the Pacific. As a concrete example, he suggested installing a terawatt's worth of nuclear power—five barge-mounted reactors of 200 gigawatts each—in Canton Island, a 6-by-9-mile atoll in the equatorial boondocks of Micronesia, some 1300 miles west of New Guinea. (A terawatt is about 769 times the electrical energy produced by the largest 1300 MWe nuclear reactors built so far, somewhere between one-seventh and one-tenth of the total power capacity of the planet!) The energy produced would be shipped as liquid hydrogen by behemoth 1600-ft long barges to population centers. Marchetti wasn't fazed by charges of technological gigantism: "The size of such reactors is certainly mindboggling to nuclear engineers, as would be the sight of a 1000 MW generator to Thomas Edison for whom a giant generator was in the range of hundreds of kilowatts," he said at the time in a paper. He still hasn't given up on the idea. He repeated his main concepts in a keynote speech at the 1998 World Hydrogen Energy Conference in Buenos Aires. According to Marchetti, a central rationale for locating giant hydrogen production centers far from civilization would be the superior transportability of gaseous fuels: electric power stations, he said, can "see" about 100 kilometers—the mean distance of electricity transport, given power consumption density patterns in industrialized nations. For a gas system with pipelines, it is about 1000 km—an area 100 times larger. Asked by an audience member whether this doesn't fly in the face of current trends such as energy decentralization, Marchetti replied, "small may be beautiful, but big is cheap after the technology has had time to settle."

3. Marchetti sums up these rules as follows: "The fractional rate at which a new commodity penetrates a market is proportional to the fraction of the market not yet covered. It includes two constants as characteristics of the particular commodity and market."

4. "Two companies develop gasoline reformers for PEM engines, on-board APUs," *Hydrogen & Fuel Cell Letter*, January 2000.

5. The "New Sunshine Program" is essentially a repackaging and streamlining of three previous programs (the Sunshine Program, the Moonlight Program and the environment technology R&D program) that Japan's government had launched in 1974, 1978, and 1989, respectively.

6. Reprint Research Report RR-97-6, International Institute for Applied Systems Analysis, 1997.

7. The 1997 Kyoto Climate Change conference—seen live on CNN, with representatives of roughly 150 nations jockeying for position and political advantage—certainly seemed to have at least some of the trappings of a Global Village town hall meeting, but world government is still too much of a stretch.

8. "Indifferent to a planet in pain," *New York Times*, September 4, 1999.

Index